THE MEDEA HYPOTHESIS

Books in the *Science Essentials* series bring cutting-edge science to a general audience. The series provides the foundation for a better understanding of the scientific and technical advances changing our world. In each volume, a prominent scientist—chosen by an advisory board of National Academy of Sciences members—conveys in clear prose the fundamental knowledge underlying a rapidly evolving field of scientific endeavor.

THE MEDEA HYPOTHESIS

Is Life on Earth Ultimately Self-Destructive?

PETER WARD

Princeton University Press   Princeton and Oxford

Requests for permission to reproduce material
from this work should be sent to
Permissions, Princeton University Press

Published by Princeton University Press,
41 William Street, Princeton, New Jersey 08540

In the United Kingdom: Princeton University Press,
6 Oxford Street, Woodstock, Oxfordshire OX20 1TW

Library of Congress Cataloging-in-Publication Data

Ward, Peter Douglas, 1949–
The medea hypothesis : is life on earth ultimately self-destructive? /
Peter Ward.
p. cm.
Includes bibliographical references and index.
ISBN 978-0-691-13075-0 (hardcover : alk. paper)   1. Extinction
(Biology)   2. Environmental geology.   3. Historical geology.
4. Life (Biology)   5. Evolution (Biology)   6. Catastrophes
(Geology)   I. Title.
QE721.2.E97W375 2009
576.8'4—dc22
2008056066

British Library Cataloging-in-Publication Data is available

This book has been composed in Adobe Caslon Pro & Trade Gothic LT

Printed on acid-free paper. ∞

press.princeton.edu

Printed in the United States of America

1 3 5 7 9 10 8 6 4 2

FOR JOE KIRSCHVINK

OUR SPECIES HAS TO TAKE CONSCIOUS ACTION ABOUT THE FUTURE OF OUR PLANET TO SURVIVE. THAT ACTION IS NOT A RETURN TO, AND RELIANCE ON, NATURAL ECOSYSTEMS, BUT RATHER SOME KIND OF TECHNOLOGICAL ENGINEERING/TERRA-FORMING TO OVERCOME THE NATURAL TENDENCY OF OUR SPHERE'S LIFE TO DRIVE ALL SPECIES, IN-CLUDING US, INTO EXTINCTION. MOTHER EARTH IS, LIKE MEDEA, THE MURDERER OF HER OWN CHILDREN, GAIA THEORY IS A FAIRY-TALE READING OF A VERY GRIM HISTORY, AND WE RELY ON "NA-TURE" TO BAIL US OUT AT OUR PERIL.

—*William Dietrich, 2006*

CONTENTS

TIME (in millions of years ago)	EON OR ERA	PERIODS (if defined)	MAJOR EARTH HISTORY EVENTS	MEDEAN EVENTS
Before 4,600	No name		Solar system (and Earth) forms	
4,600–3,800	Hadean		Origin of Earth to origin of life	
3,800–2,500	Archean		First life to first eukaryotes	Great oxygenation event; first Snowball Earth
2500–543	Proterozoic		First eukaryotic cells to first skeletonized animals	Multiple Snowball Earth events; mass extinction of Ediacarans
543–250	Paleozoic	Permian Carboniferous Devonian Silurian Ordovician Cambrian	First skeletonized animals to Permian mass extinction	Life causes mass extinctions at end of of Cambrian, in Devonian, and at end of Permian; colonization of land by plants causes major ice age
250–65	Mesozoic	Cretaceous Jurassic Triassic	First dinosaurs to Cretaceous mass extinction	Life causes mass extinctions at end of Triassic, in Jurassic, at end of Jurassic, and in Cretaceous
65–present	Cenozoic	Neogene Paleogene	First large mammals to end of terrestrial communities; during this time ice age ends	Life causes mass extinction at end of Paleocene; life causes long-term cooling and loss of planetwide forests; life causes Pleistocene ice age; humans evolve

INTRODUCTION

Let us begin with a thought experiment. Envision the vast, life-filled rain forest now occupying the Amazon Basin of modern day Brazil. The wide, brown river slowly but inexorably flows eastward, carrying within its fluvial grasp unnumbered tons of mud, silt, sand, and in some places even gravel, originating either in the foothills of the rapidly eroding Andes Mountains far to the west, or from the upper reaches and banks of the river itself. Commingled with this future sedimentary rock are vast quantities of rotting plant material, ranging in size from entire trees to microscopic fragments of peat. This produce is fed upon by armadas of herbivores, stalked in turn by carnivores, with scavengers patiently waiting for both to become a meal. Vast, jungle forests grow right to the water's edge, made up of trees reaching upward toward an unseen sky, their highest leaves finally emerging from the multiple layers of canopied gloom to form a vast plateau looking like a green sea swaying gently in sync with the breezes under the bright equatorial sun, a variegated surface far above the twilight world of the forest floor. Amid and above this canopy flit insects, birds, and bats, an aerial nekton among the simpler airborne plankton that, combined, turn the sky darker at dawn and dusk by their numbers and fill it with sound at any hour. This Amazon Basin with its attendant rain forests is a cornucopia of multiplying and rotting cells, a place of rapid growth and rapid decomposition, a habitat inhabited at a frenetic pace by an unknowable biomass of life, an unknowable level of species diversity, and a barely understood morphospace of body plan disparity.

The Amazon Basin, this storehouse of diversity and biomass, is not unique on our planet. We have only to continue eastward from the merging of the Amazon with the sea, following the equatorial latitude as a guide across the South Atlantic Ocean until we are over land again, over African rain forest this time, once again over life exorbitant in both kind and numbers. As in South America, this jungle spans the entire width of the continent we pass over, ever eastward, until we are over water again—this time the Indian Ocean. When next we reach land, we are over the South Asian rain forests.

The species change in each of these disparate rain forests, but the sheer abundance and exuberance of life in these habitats do not. These hot rain forests are the most populated and most diverse—two different things—storehouses of life on planet Earth. More species are seemingly crammed into the trees alone—perhaps more species of *beetles* live in the trees alone—than there are other species in all the other habitats on Earth combined, including the coral reefs and surface regions of the sea. We are talking as many as 30 million beetles in addition to everything else! But there is so much else. If we were to weigh the total mass of living material from these forests—from microbes in the soil beneath to the tiniest of float spiders in the air above—the weight of this living material would likely be a significant (but impossible to really fathom) fraction of the Earth's total biomass, perhaps even exceeding that of all of the rest of life combined—again including the coral reefs, the plankton on the surface of the sea, the cool boreal forests and wheat fields, and the deep microbial biosphere.

And now for our experiment. Rolling up our sleeves, godlike, we change the climate over these forests by cooling the Earth's temperature just a few degrees. A very simple act, but one that quickly takes on a life (death, actually) of its own, for the cooling begets more cooling in a "positive" feedback mechanism: the cooler it gets, the more the Earth cools. As the Earth cools, the poles become more snow-covered, and more light is reflected back into space, causing the temperature to cool some more, and more ice to form. In the tropical forests there is no ice, of course, nor will there ever be, but the temperature changes have caused a decrease in and, more important, an irregularity of rainfall; areas with monsoons lose them, while other areas accustomed to constant, rather than episodic, rain experience the first dry and wet seasons. Those plants needing year-round rainfall begin to die, and those animal species confined to these particular trees in some obligate fashion die too.

To the north and south, the ice twins of oceanic ice caps and continental glaciers emerge from sleep and, like vast, stretching giants, extend their fingers toward one another until they merge into white wastes of ice, over a mile thick on land, and once thus united they swiftly move either south or north depending on their starting points in the high latitudes of both the northern and southern halves of the Earth. In both hemispheres they scrape their way toward those warmer regions of the midplanet, sucking the level of the sea down to feed their growth as greedy solid water changes phase from its more temperate liquid brother. Their continent-wide fronts are heralded by powerful and dreadfully cold adiabatic winds that blow massive piles of dust onto formerly fertile lands, and even the lands beyond the reach of the glaciers and winds are affected. No place on Earth escapes a change of its local climate in this new, reverse Oreo cookie of a world, all white top and bottom, with a dying and drying, darker middle of plants and rock, for no place keeps the exact or, in most places, even approximately the same kind of climate that existed in the preglacial world.

In the non–ice covered equatorial latitudes we watch as many tree species begin to die, bringing down the complex food webs that are dependent on the trees and their ecological cornices, webs far more complex than any spider's weaving found stretched between the branches of the closely spaced limbs. In the rain forests and dry forests alike, trees swiftly die and fall, to be replaced by grass and weeds at best, and bare rock elsewhere, their encompassed species of animals and plants changing from living to extinct, with virtually no possibility even of a fossil record to mark their passing.

The process of deathly change does not stay restricted to land. As the land plants die, the rivers become choked with rotting vegetation that makes its way out to sea, causing a short-term bloom of life amid the newly abundant nutrients. But soon that bloom ends as oxygen is used up, and the rotting vegetation falls onto oxygen-free bottoms, creating a vast eutrophication, where so much organic material rots that it takes all of the available oxygen out of the surrounding water. The death of trees on land has more consequences: with roots gone, the soil is carried away by wind and rain, eventually falling onto the ocean bottoms. Near the continents, great submarine fans are built, covering the once stable bottom communities, choking the sea bottom. The more mobile invertebrates dig out of the less catastrophic, underwater landslide events, but the immobile fauna has no chance, and in the abundant cases of the larger undersea landslide flows, there is no chance at all. All are buried by layers of sediment rumbling off the land's edge into the sea. The river mouths choke; rapid cover of sand and silt kills off the fragile estuarine communities as deltas and rivers radically change course in the mounting piles of sediment streaming off the land.

There is no refuge as the temperatures of the sea change, driven downward by the cooling atmosphere. Species adapted to warmth are killed directly, and as these die they doom the many more species dependent on stable food webs and predicted resources. The extinction on land thus begets its own, different kind of misery in the sea. Thus we have no need to wave our evil magic wand over the seas; simply killing off the land forests, or a good proportion of them anyway, does the trick.

The increasing ice on land is ultimately composed of seawater, and so much ice forms on land that the level of the sea in its many oceans precipitously falls. A consequence of this is that the falling sea level exposes near-shore, formerly undersea communities. Eventually the sea drops nearly 250 feet below present-day levels, killing off the richest undersea communities till that time still extant—including the coral reefs, which have to migrate seaward to avoid being stranded, but which become stranded over vast regions nevertheless. The entire Great Barrier Reef tract of Australia dries to salt marsh, and the Torres Strait now separating New Guinea from Australia dries up as well.

But there is far more to come. The Earth's climate has long been on a knife-blade balance between times of glaciers and times without, and our push to just slightly colder temperatures has tipped that scale: the glaciers have now grown well down into the mid-latitudes of our planet, scouring to glacial flour whatever plant communities that are left, stomped under a mile of ice. Armadas of icebergs calve off the glacial fronts where they reach the sea, dropping piles of gravel far out to sea as they melt. The atmosphere is perpetually hazy with dust.

We can now return to where we started. Following our catastrophic meddling, most of the Amazon is now grassland and savannah, with only isolated pockets of the once lush rain forests. Large areas are bare rock, since the soil beneath these forests had been thin to begin with. The changing weather patterns and death of trees stripped many regions into Depression era dustbowls; life did not have the tens of thousands of years necessary to change that bare rock back into productive soil.

So far, our experiment has been on a world where there are no humans. But that is not our world. Add our civilization into the mix and make a monetary calculation of what will be lost: all coastal cites are now perched far from the sea; the dust is so thick that jet engines clog in the stratosphere, and we are reduced to propeller plane travel; the changing weather patterns and the cold have wrecked any *Farmers Almanac*, and most agribusiness as well; large proportions of the formerly rich wheat and corn regions of the United States, Canada, northern Europe, and the ex-Soviet Union are changed, with some showing increasing harvest, but most yielding less. And it is not just wheat that takes a hit: many of the rice-growing areas of China are either under ice or so near the glaciers with their high winds and dust that they are no longer productive. What would it cost to rebuild all coastal cities, to replant most crops in new places, to fight an inevitable famine and fight as well the border wars precipitated by the many millions of humans who are displaced by the economic ravages inflicted by the falling sea, displaced into entirely different countries? The cost would be in the trillions of dollars. And this is just an accounting of the human constructions lost. What is the monetary value of a species going extinct?

Let us try to tally up what we have done beyond the monetary aspects. We can do this in two ways: the number of species going extinct, and the relative biomass of the planet, before and after. Both are as yet difficult to put final numbers on, but our science does tell us that both will be significantly large.

As we stare out at what we have wrought, we might wonder at the penalty for such death bestowed. What should such a killer receive as penance, when a million or more species, a significant portion of the total amount of life on Earth, are killed off as a result of some action that changed a world with planet-spanning rain forests to one of desert, ice, and grassy plains, and but a thread of the once luxurious, pole-to-pole forests? In China, according to (perhaps) urban legend, the perpetrator of any crime costing the state a significant loss of money is summarily executed, with the tab for the bullet to be paid by the miscreant's family. What punishment is just for the perpetrator of a crime vastly more immense in scale?

We can rejoice that this particular thought experiment is just that—a thought experiment. But in fact such global cooling events have been triggered in the past by deadly murderers. And cold is only one of their weapons. If the global assassins wanted to kill off an even larger percentage of life on Earth than is possible with glacial cold, the weapon of mass destruction of choice would be heat, or, more accurately, that now familiar phrase "global warming." By just raising the temperature of the planet a few more degrees than it is now, we could reduce Earth-life's diversity and biomass to extremely low levels—or even cause planetary sterility through mass extinction.

Bogeymen in the closet can make for a good scare. The shiver of a slasher movie is a cheap thrill soon ended as the fictional killer is put to its deserved end, the lights go on, the movie-goers file out, and we go on with our lives. But the reality is that there *is* a killer on the loose capable of planetary-scale catastrophe, a killer quite alive, an assassin that owes nothing to Hollywood or Stephen King. It has killed in the deep and near past and is poised to kill again in the near and distant future. This killer is devilishly sly, cleverer than any well-disguised and innocuous character thought up by the pulp mystery writers. It hides in plain sight. The purpose of this book is to put a face on this, life's worst enemy.

The killer is life itself. If left unchecked, it will hasten the ultimate death of all life on Earth. Only human intelligence and engineering can delay this fate.

Hasn't ended life in 5 billion yrs!

OR soc!?

· · ·

There is nothing conscious about life's lethal activities. In fact, the individuals making up each species are ruled by natural selection and live in ways maximizing their survivability, and the repeated killing off of many of them is definitely not in their interest. But, perhaps paradoxically, aggregates of species interacting with the physical environment as well as with other life, in the region on Earth that we term the biosphere, appear to have effects not selected for—lethal effects, in fact.

We humans have the odd distinction of being the only ones that either know or care, and it remains to be seen if we will stave off planetary extinction or hasten its onset. Right now we are rapidly transforming a world of moderate temperature (one with ice caps and relatively low sea level) to a heated world without ice caps and with high sea level. Yet while we are the only organisms capable of extending the life of the biosphere, we are certainly not the only creatures capable of the opposite—reducing the lifespan of the Earth as a habitable world. The mass extinctions of the past were far more lethal than any war yet waged by humans, and they were not even primarily caused by complex life. Rather, microbes were at their root. But higher life forms were accessories to this murder, in that it was higher life that allowed microbes to multiply to the point that they could begin their ruthless poisoning of air and sea. What we must do is to defang the monster through direct intervention into the carbon, sulfur, nitrogen, and sulfur cycles, and by maintaining planetary temperature so that there are always ice caps in the high latitudes. Unfortunately, many influential and well-meaning "environmentalists" urge humanity to give control of the world back to "nature," to return things as much as possible to how they were before we evolved. A second purpose of this book is to show that such an action would be suicide.

End of life?!

For who?

In Saturday matinee parlance, our planet is a jet plane in the hands of a madman bent on spectacular immolation of self-destruction. There is only one chance of survival: human intervention/engineering. We must seize the controls of the various elemental cycles that determine the fate of life on Earth and pull up out of this fatal, albeit slow (for there is no other way to describe hundreds of millions of years) dive. And we must realize that life does not optimize the world for itself, as well as accept the fact that we live on a rapidly dying planet; our ultimate "Earthlike" planet, so beloved by the new science of astrobiology, has been brought into its old age by life itself.

What kind of Mother Nature would do such a thing? Should she even be called a "Mother"? Surely we cannot consider her a *good* mother.

Just how and why, did Mother Nature get so vicious? The answer seems to be that an innate lethality is a side effect of the main factor leading to life itself—the process that we call evolution, in reality a complex of forces that first brought life into existence and then shaped it, divided it, and spread it throughout the biosphere. But a characteristic of evolution is that its basic unit is the species, not the biosphere, and from this accrues a vicious, uncaring lethality toward other species that is one of the three most basic characteristics of life itself.

One of the great discoveries of the biological and geological sciences has been the realization of how important life is in affecting its own habitability (livability). We now know that Earth life has always had a major effect on the nonliving parts of the Earth where life lives, and that we expect it always will—from the formation of land forms and biotically constructed structures such as reefs and forests, to the composition of the atmosphere and chemistry of the oceans. And beyond the nonliving, life has profound effects on itself—that too is clear. From the spring floods of terrestrial nutrients that trigger oceanic plankton blooms, to the change of the Earth's temperature itself through the effects of plant cover on the Earth's albedo, life affects life. But the other side of this coin seems true as well: life affects death, and, as I will attempt to show, the ultimate death, the end of living organisms on Earth, will be dictated by life well before external factors working toward the same end (an enlarging Sun, causing a loss of oceans, the loss of atmospheric oxygen, and a lethal increase in planetary temperature) can operate.

I understand that this particular view of life—that (to put it mildly) life is less than benign to species other than itself—is a minority view. There is a vast library of articles and books essentially suggesting (unlike the dark view of life described above) that evolution works most basically on the biosphere rather than at the level of species and in so doing has optimized planetary conditions since its inception, allowing ever-greater masses of life to exist. This has happened through a series of geological, chemical, and biological "feedback" systems, which have acted as checks and balances on those conditions that affect life (such as temperature, pH, and atmospheric composition) in ways that have kept the planet suitable for life's existence, and which might even have enhanced conditions for life. There is even a large body of scientific literature claiming that, ultimately, life will extend the life of the biosphere beyond what physical conditions would dictate. The planet gets hotter because of a more energetic Sun? Life cools the planet by enhancing chemical weathering. Life is threatened because there is too little usable carbon in the atmosphere for photosynthesis? Life evolves new methods of carbon acquisition. There is too little/too much of one of the elements necessary for life, such as carbon, oxygen, sulfur, phosphorus, or nitrogen? Life evolves new ways of metabolizing that change (increase/decrease) the availability of these elements. Thus altering the physical aspects of the Earth to increase habitability could range from increasing the amount of nutrients available, to controlling the temperature and atmosphere of the planet to the extent that those conditions, if not optimal, would at least never stray into antibiotic extremes.

These are thought-provoking scientific conclusions. But to what extent are they true—if at all?

We are faced with two very different (albeit linked) hypotheses. First, has life changed the physical aspects of the Earth in such a way as to either maintain or even increase habitability of the planet—and thus increased the diversity and/or biomass of Earth life on a planetary scale over and above what it might have been if only physical conditions were in play? Second, through such (or other) actions, will life extend—or shorten—the biosphere's ultimate longevity (the time during which life will continue to exist on the planet)? Answering the latter question requires us to find out to what extent the future of life on Earth will be determined by external (nonbiological) factors such as an enlarging Sun—compared to the effect of life on itself and the various systems allowing life to exist on this planet at all—and if life can somehow mitigate or ameliorate physical conditions adversely affecting itself.

The stakes are higher than just understanding the nature and fate of Earth life. Who of us imagines that, of all the nearly innumerable planets and moons in the cosmos, ours was the only one blessed (or cursed?) with life? Can the answers to the questions above yield insights not only about our own planet, but for what must be untold numbers of other inhabited planets, whose inhabitants surely range from the equivalents of microbes to intelligences equal to or exceeding our own?

How to proceed? The scientifically observed past and modeled future will here serve to test two very different hypotheses about the effect of life on itself, including its future on Earth. The first, known as the *Gaia hypothesis*, is actually composed of at least two separate hypotheses, including one called "Self-regulating Gaia" and another called "Optimizing Gaia." Each of these will be defined in detail in a later chapter, but both can be said to support an overarching hypothesis stating that Earth life in the past, present, and surely the future has had and will have the effect of maintaining planetary habitability by affecting the external environment in such a way as to keep it within specific limits dictated by various tolerances and requirements of life. The more extreme form of this hypothesis (Optimizing Gaia) says that not only has and does life maintain "habitability" for itself (albeit unconsciously, simply as an inherent property of itself), but it actually improves conditions by changing such factors as planetary atmospheric and oceanic chemistry, the cycling of elements through the biosphere, and the availability of nutrients to levels *more favor-*

able for life. Finally, there is an even more extreme interpretation, one not backed by science or scientists, but one that seems to have entered the popular consciousness—that the Earth itself is an actual living entity: a living organism orbiting our Sun. That is the bad side of things. But an undeniable offshoot of the Gaia questions has been the birth of an entirely new branch of science called Earth System Science, and this new field has proven both vigorous and enormously rich in the contributions of its scientific discoveries. It has attracted some of the best brains in all of science, and thus a huge debt is owed to those who originally introduced the Gaia hypotheses to the world.

Some of the various Gaia hypotheses are rapidly approaching a half century in age, while several others are already decades old. To some of their scientific authors, these hypotheses have passed a sufficient number of scientific tests that they can be united and elevated to the much stronger scientific level of a theory, rather than simply the more equivocal level of a hypothesis. Many other scientists disagree, however, and for decades there has been spirited discussion both pro and con. Yet while many interested scientists have scientifically tested various aspects of the Gaia hypotheses, with some of that number finding them wanting (and thus rejecting them), this "anti-Gaian" contingent has never offered a competing hypothesis. Here, one is proposed. The rationale for this is simple: hypotheses demand tests, and scientific tests spur progress. Furthermore, a large number of rather new discoveries about life of the past as well as new models about life in the future in my opinion readily falsify all of the Gaia hypotheses. Science abhors a vacuum.

The name Gaia means "Good Mother," and the Greeks referred to her as Goddess of the Earth. I hypothesize that life and its processes, together often referred to as "Mother Nature," was, is, and will be anything but a good mother to her many evolved and evolving species. So here (only semi-jocularly) I propose the "Medea hypothesis," named after one of the worst mothers ever, as an alternative to "Mother Nature."

Medea, daughter of the king of long ago Colchis (modern day Georgia, on the Black Sea), married the famous Argonaut, the captain of the ship *Argo*, Jason—the Jason who, in legend, stole not only Golden Fleece from Aetes, the king of the Colchians, but ran off with his daughter to boot. (Jason could cast a spell over any woman. Medea really didn't like him—Jason apparently wasn't very likeable—but she apparently had no control over the matter.). Following much carnage, subterfuge, boat chases, and long journeys to purify various sins and put to rest unhappy ghosts, Jason, Medea, and their nautical crew the Argonauts returned to Greece with the Fleece. There, Medea bore Jason children, but she soon found Jason to be not the man she was enchanted to believe that he was, and in an act of rage (for Jason was as bad a husband as she turned out to be a mother) she killed all of their children. This name thus seems appropriate for an interpretation of Earth life, which collectively has shown itself through many past episodes in deep time to the recent past, as well as in current behavior, to be inherently selfish and ultimately biocidal. A result of this bad mothering, I propose, will be a shortening of the time that life will exist on our planet. Life will do this to itself by unconsciously changing environmental conditions to a point where there can no longer be plant life or, ultimately, any kind of life.

To argue my case, I will use new discoveries from geology, biology, and most of the fossil record. To me, these new understandings are like a memory exhumed from some deep sleep, in reality from the deep past, that shows the absolute need to construct a new paradigm about both past and future, one that will require a rather painful shift from the kinds of conservation and environmentalism that are practiced now. The philosophical underpinnings of modern environmentalism are that the planet must be returned to environmental conditions that existed prior to the evolution of humankind's technological civilization, with the resulting planetwide changes to almost every facet of the environment. Instead, we humans must resort to wholesale planetary engineering if we are to overcome the tendencies of life around us—and those of our own species—to make the Earth a less salubrious (and eventually lethal) abode for life. The sum of this record, which is meant to be the theme of this work, is the interpretation that *the evolution of life triggered a series of disasters that are inimical to life and will continue to do so into the future.*

If true, one implication is that the environmental challenges confronting our species and its civilizations are far more than simple overpopulation and all that entails. The fact is that we live on a rapidly aging planet, and we will soon have but two choices if our species is to survive: engineer on a planetary scale or get off. Instead of restoring our planet to how it was before humans, we have to do exactly what the Gaia hypothesis suggests that life has done all along: optimize conditions for further life. We have to confront the nature of life itself and deal especially with groups of life that we animals have battled throughout our history: armies of microbes that cause their own kind of pollution, inimical to our kind of life.

if true, how?

I will try to show in the pages to come that the cause of this inherent tendency of life on Earth is due to one of Earth life's most deeply inherent characteristics, so deeply rooted that it would not be life without this aspect. It is that all Earth life is a slave to a process called evolution, Darwinian evolution in fact, for Charles Darwin got the process spectacularly correct even without understanding how any characteristic could be heritable. Along with replication and metabolism, evolution is one of the three tripods that defines life on Earth; take any of these legs away and it falls into the nonlife category. Life can no more help evolving than we can stop breathing and stay alive. You evolve or your species goes extinct, for the Earth keeps changing, and the formation of our own form of life was made possible because of this characteristic. When life first appeared, some 3.7 billion years ago at the latest, our planet was a far more energetic and dangerous place to live on or in, and only through the ability to change generation by generation could the earliest forms of life survive. It was not only survival of the fittest, but also survival of the best and fastest evolvers. Natural selection not only worked on better ways to get energy and withstand environmental difficulties but evolved better ways to evolve. Before all else, life worked on perfecting energy acquisition, replicating quickly and with fidelity, and evolving ever more quickly. But the price to pay is that each and every species innately "tries" to become the dominant species on the planet, with no regard to other species. Be it bacteria or bees, all try to produce as many individuals as possible and in so doing can and do poison the environment in various ways for all other species, including the species in question.

How much longer will the Earth sustain life in the face of this relentless overpopulation by a variety of species, which tends to use up resources—unless we humans step in and save things, of course? Alone among all the creatures large and small, our species can extend the length of the biosphere on Earth, which, like all of us, has a finite lifespan. Yet that lifespan, currently dictated by life itself, can be lengthened. Vastly lengthened.

I address these issues in the following way. Chapter 1 defines life, and then Darwinian life, which may be a subset of life. Chapter 2 discusses what "success" means to life, and chapter 3 examines two different and mutually exclusive hypotheses about one of the most fundamental aspects of life: does life improve a habitable planet for more life, or does it reduce habitability? Chapter 4 looks at planetary "feedbacks," chapter 5 is about a succession of events in deep time giving evidence and allowing us to choose between these two hypotheses, chapter 6 looks at humans as Medean forces, and chapters 7 and 8 consider biomass of the past and future. Chapter 9 provides a summation of the scientific evidence and allows us to choose among competing hypotheses, while chapter 10 looks at the societal implications of that choice. Finally chapter 11 takes a longer view, describing solutions to problems brought up through the book. It deals with engineering and technological issues for both extending the life of the biosphere and possibly escaping a dying Earth to somewhere else in the cosmos.

Introducing this new hypothesis puts up a second piñata for scientists to joust with, and this hypothesis hangs right next to Gaia. Some might find this new view of life depressing. I find it exhilarating, for if correct, only we humans (or other intelligent species out there in the cosmos) can change the rules and save the rest of life, as well as our own species, from itself.

THE MEDEA HYPOTHESIS

1

DARWINIAN LIFE

> From so simple a beginning endless forms most
> beautiful and most wonderful have been, and
> are being evolved.
>
> —Charles Darwin, *On the Origin of Species*, 1859

In the summer of 2007 I entered into a new experience: teaching the science of evolution to entering university students. Each of the nineteen students in my class, none older than eighteen years of age, started his or her first university class with some mixture of optimism and trepidation. Most, it turned out, wanted to be scientists. Yet in a series of short papers, most also readily admitted that while they had been well prepared in their high school classes in various mixtures of mathematics, chemistry, physics, and biology, virtually none had learned anything about what is variously labeled as evolutionary theory or, if one has a more creationist bent, Darwinism.

The reason for this omission was easily ascertained. Most high school teachers have stressful enough lives dealing with the daily traumas of teaching in U.S. high schools—why add extra drama by entering into one of the most emotionally charged of all subjects, evolution? Many a teacher has had the very unpleasant experience of describing theories about human phylogeny and meeting an angry, fundamentalist parent soon thereafter. So the subject is largely ignored.

Unfortunately, when evolution is ignored, other allied sciences are ignored as well. Perhaps the most important of these deals with the origin of life. For reasons also obscure, one of the most perplexing and important of scientific questions—how life first appeared on Earth—is ignored in basic university biology courses and, if mentioned at all, is discussed in brief detail in a more advanced evolution course. But this seems curious, for how life first appeared and how it later evolved the ability to evolve were different processes. (As we shall see, the ability to evolve became an inherent property of Earth life, but surely only after the synthesis of life's building blocks.) The current modes of evolution involving genes on DNA, itself massed together in a chromosome, were a long way in the future when various snippets of amino acids were assembling into some proto-RNA molecule. Yet how life first came into existence is a viable field of study, and for want of any better place this topic is usually dealt with in an evolution class.

Thus, on my second day in class, I asked the assembled multitude to write me a short essay on the definition of life. The results were all over the map. While some honed in on chemical definitions, the majority leapt toward the metaphysical, imbuing life with a vast array of mystical properties, ranging from the minimal to truly godlike. What came through, however, was a fundamental property that does not usually make it into the textbooks but is at the heart of the arguments here: that life in the aggregate acts very differently from life as an individual. However, those imbuing life with properties over and above those of an individual saw those properties as inherently "good" and helpful to other life—with the sole exception of we humans, which were viewed rather guiltily as not following life's lead in making things better. I agree with my students' prescient sense that life as a whole acts differently from life as an individual. Where I disagree is about the ultimate effect of life on itself.

An analogy about this disconnect can be seen in the relationship between individual humans and the human race. Each of us lives our life, usually hoping for, and living in ways to create, as much happiness as possible. Many of us work diligently to reduce the environmental "footprint" of our existence. Yet in aggregate we are clearly changing the physical Earth, and changing conditions for both ourselves and other life. So too with "life": as an aggregate it has major effects on itself as well as the planet. This aspect of life might certainly help explain some of the behaviors proposed at the heart of my arguments here.

Let us begin this argument with the most minimal definition of life, followed by a definition of Earth life, for it is one of life's inherent properties that is the heart of the problem.

The question "What is life?" is deceptively simple, with no simple answer. Perhaps the most parsimonious answer is as follows: "All life forms are composed of molecules that are not themselves alive." This definition, as imprecise as it is, does hint at a deeper truth. At what level of organization does life "kick in"; in what ways do living and nonliving matter differ? Most who have thought deeply about what life could be, and what chemical forms it could take (Ward 2005), *believe* that Earth life is but one kind of possible life. But no one on this Earth can prove that there is any life beyond that of the Earth, and indeed one of the astonishments about life on Earth is not how diverse it is (which of course it is, at least at the level of species), but how *poor* the Earth is in the kinds of life. While those who worry about biodiversity rightly point out how the Earth is losing species, the reality is that there is only one kind of life on Earth—our familiar DNA/RNA life. E. O. Wilson's magnificent 1994 book, *The Diversity of Life*, could in reality be retitled *One*.

But what are the characteristics of life, and then Earth life, and why are these central to the arguments here? I will argue that one of the attributes that most experts equate with being "alive" is the ability for the entity to evolve in a way that would have been familiar to Charles Darwin, an evolution that now bears his name: Darwinian evolution. It is that aspect of Earth life (and perhaps all life, since the very number of stars in the cosmos make the presence of life beyond Earth as close to a certainty without being one as there could probably be) that, to many, is the source of Earth life's singular success. And yet how important is the behavior of an individual compared to the behavior—or, perhaps more properly, the effects—of the collective? My thesis is that the inherent property to evolve is also the source of the inherent "suicidalness" of life—a facet of what I will define as the *Medea principle*, to be posed and referred to here as a hypothesis.

3

Perhaps a better question than "What is life?" is "What does life do?" Physicist Paul Davies, who has pondered the "What is life?" question more than virtually any other thinker, listed the following:

> *Life metabolizes.* All organisms process chemicals and in so doing bring energy into their bodies. But of what use is this energy? The processing and liberation of energy by an organism is what we call metabolism, and that is necessary to maintain internal order.
>
> *Life has complexity and organization.* There is no really simple life, composed of but a handful of (or even a few million) atoms. All life is composed of a great number of atoms arranged in intricate ways. But complexity is not enough; it is organization of this complexity that is a hallmark of life. Complexity is not a machine. It is a property. It is also something the life "does."
>
> *Life reproduces.* This one is obvious, and one could argue that a series of machines could be programmed to reproduce, but Davies makes the point that life must not only make a copy of itself, but also make a copy of the mechanism that allows further copying; as Davies puts it, life must include a copy of the replication apparatus too. Again, there are machines that allow life to copy itself, but the process is not that of a machine.
>
> *Life develops.* Once a copy is made, life continues to change; this can be called development. Again, this is a process mediated by the machines of life, but also involving processes that are quite un-machinelike. Machines do not grow, nor change in shape and even in function with that growth.
>
> *Life is autonomous.* This one might be the toughest to define, yet it is central to being alive. An organism is autonomous, or has self-determination. But how "autonomy" is derived from the many parts and workings of an organism is still a mystery, according to Davies. Yet it is that autonomy that again separates life from machine.
>
> Finally, Davies noted: *life evolves.* According to Davies, this is one of the most fundamental properties of life, and one that

[handwritten margin note: Wittgenstein (?)]

is integral to its existence. Davies describes this characteristic as the paradox of permanence and change. Genes must replicate, and if they cannot do so with great regularity, the organism will die. Yet, on the other hand, if the replication is perfect, there will be no variability, no way that evolution through natural selection can take place. Evolution is the key to adaptation, and without adaptation there can be no life. Again, a process, not a machine.

Davies is far from alone in advocating that Darwinian evolution is a fundamental property of life, nor was he the first to do so. A decade before Davies so eloquently made these observations about life, the great Carl Sagan famously wrestled with the question of what life is. Unlike most others thinking about this topic, who were dealing only with life as it is found on Earth, Sagan came at the problem with a specific goal: he was interested in life beyond Earth, and at the time of his observations about life, in the mid-1970s, he was involved in several NASA missions involved in searching for such life, most famously the Viking missions to Mars. Sagan's definition of life, which was largely taken up by NASA and is still used to this day, sees life as *a chemical system capable of Darwinian evolution*, meaning that there are more individuals present in the environment than there is energy available, so some will die. Those who survive do so because they carry advantageous heritable traits that they pass on to their descendents, thus lending the offspring greater ability to survive.

The view that evolution is an inherent property of life has come to be called the evolutionist view. For instance, life has been defined as being a self-sustained chemical system capable of undergoing Darwinian evolution, as well as a self-replicating, evolving system based on organic chemistry, as well as a system capable of evolution by natural selection. Finally, life has been called a material system that undergoes Darwinian evolution.

So just what is "Darwinian evolution"? We should briefly describe its basic tenets before going any further. While Darwin is credited with a "theory of evolution," in fact he proposed two separate and

5

Huh?
No he
didn't
Not critical
but annoying

testable hypotheses. The first is that all life on Earth came from a single common ancestor. Second, he proposed a principle of variation: that life reproduces to produce slightly different variants of the parent (as well as progeny that closely match the parent, or, in reproduction through cloning, forms that are genetically similar). But Darwin also noted that most "parents" produce more offspring than can live because of shortages of food, or shelter, or other necessities of life. Because of a surplus of offspring, in most instances some will perish. Those that survived did so in the long run—and indeed we are talking of many generations—because they had characteristics that made them in some way superior to others of their own species. These characteristics, such as larger size, which is a very common trend in evolutionary lineages, must also be "heritable"—that is, the characteristics have to be passed on to the next generation.

Darwin saw this competition as "survival of the fittest," and he gave the process the technical name "natural selection." Over the long run, the survivors would be those with characteristics (hereditable characteristics, that is, ones that can be passed on to the next generation and not just ones acquired during the life of the individual, such as a human sex change operation) lending the greatest "fitness," or ability to survive. Examples are many, such as the few giraffes with ten-foot-long necks among a herd with seven-foot necks in places where the lowest vegetation is nine feet above the ground; the fastest-swimming fish in a lake where the predators can catch the slowest and even median-velocity swimmers. These survivors then pass on these successful characters to their offspring, and evolution has taken place.

Over simplified
simplistic
but ok

Speciation, the formation of an entirely new species, is a larger-scale process. A species is deemed separate if it can no longer interbreed with its parent's populations. For new species to form, most commonly there must be geographic isolation of a subset of a smaller population into a new environment cut off from the larger population, one that has different challenges for survival. Over some generations these new environmental challenges would cause evolution of forms dissimilar enough that if the two populations should again come into contact, the two gene pools are now so different that breeding does not produce successful offspring.

allopatric
vs
sympatric

Thinkers on the subject have come to agree that Darwinian evolution is certainly a key property of Earth life, or RNA/DNA life, and perhaps it is a necessary property of all life in the Cosmos.

DEFINING EARTH LIFE

With all the apparent diversity of life on Earth, all Earth life yet discovered shows a unifying characteristic—it all contains DNA. This is why I suggest that the true diversity of life on Earth is 1.

not virus or prion?

Composed of two backbones (the famous "double helix" described by its discoverers, James Watson and Francis Crick), DNA is the information storage system of life itself—the "software" that runs all of Earth life's hardware. These two spirals are bound together by a series of projections, like steps on a ladder, made up of the distinctive DNA "bases," or base pairs: adenine, cytosine, guanine, and thymine. The term "base pair" comes from the fact that the bases always join up: cytosine always pairs with guanine, and thymine always joins with adenine. The order of base pairs supplies the language of life: these are the genes that code for all information about a particular life form.

If DNA is the information carrier, a single-stranded variant called RNA is its slave, a molecule that translates information into action—or in life's case, into the actual production of proteins. RNA molecules are similar to DNA in having a helix and bases. But they differ in usually (but not always) having but a single strand, or helix, rather than the double helix of DNA. Also, RNA has one different base from DNA.

RNA is tantalizing stuff. While indeed it is "hardware" in carrying amino acids to protein-building sites in the ribosomes, it is clear that some RNA has multiple functions, including information storage. There is an apparently important regulatory role played in eukaryotes by nonprotein-coding RNA, which is an example of RNA acting simultaneously as software and hardware.

DNA provided the answers to many of the mysteries of genetics, answering the question, once and for all, about what is a gene, for the nature of inheritance, from Darwin's time to the twentieth century, had remained a most vexing question. James Watson and Francis Crick made the great discovery—one that launched an enormous

No – but ok

7

revolution in biology—and their great discovery was announced in a paper in the journal *Nature* that was but a single page long. Their finding was actually a model, not an experimental result, but the model had enormous predictive power. It became clear that a gene is made of DNA, and that one gene makes one protein. Watson and Crick proposed that one half of the DNA ladder serves as a template for re-creating the other half during replication. Each gene is a discrete sequence of DNA nucleotides, with each "word" in the genetic code being three letters long.

How does a gene specify the production of an enzyme? It was Francis Crick who suggested that the sequence of bases was a code—the so-called Genetic Code—that somehow provided information for the formation of proteins, one amino acid at a time. The information coded had to be read (transcribed) and then translated into proteins. That is where RNA comes in. Earth life uses twenty amino acids. Not nineteen. Not twenty-one. *And always the same twenty!* DNA codes for RNA, which codes for proteins, which are all made up of combinations of the twenty amino acids. This, then, the central dogma of molecular biology, may also be called a central characteristic of Earth life.

HOW EVOLUTION ARISES

Genes are the blueprints necessary to make Earth life's major structural and chemical partner: proteins. Proteins perform the various functions of the cell. A protein's action is determined both by its chemical constituents and by its shape. Proteins become folded in highly complicated topographies, and often their final three dimensions shape their actions.

So how does DNA specify a particular protein? A typical protein might be made up of 100 to more than 500 individual amino acids (but all of those same twenty kinds), and thus its gene, the sequence of nucleotides coding for the protein on the DNA strand (since the string of amino acids that make up the protein are coded on the DNA strand), will be composed of 300 to 1,500 or more sets of "steps" on the DNA ladder. These are arranged in linear order along the DNA strand, like letters in a sentence. And, like a sentence, there will be spaces and punctuation as well (like *stop!*). The RNA slaves grab these and take them to a ribosome, where the actual protein is constructed.

This information flow mainly goes one way only—from DNA to RNA (though, as noted above, some RNA has no role in protein formation but functions as a regulatory molecule). The poor RNAs have no say in any of this: go here; build that, bossed forever from above by DNA. All the proteins being built by the ribosomes, at the direction of the RNAs (themselves slaves to the DNA), do one of two things: they build a structure, or, more commonly, they function as enzymes that catalyze a chemical reaction in the cell itself important for maintaining life function—such as metabolism.

Having a DNA is obviously not all there is to life. We need a wall (membrane) to enclose our cell, and a solvent to fill it with. Both the wall or membrane structure and the solvent are also features that we can use to identify common Earth life. The biochemist Steven Benner also suggests that a requirement of life is some sort of scaffolding, for both building blocks of our life structure and to hold biomolecules in correct orientation so as to allow chemical processes of life. Our Earth life uses carbon as the scaffolding element, but silicon could be used as well if there are side branches on long chain carbons on which silicon compounds could bond.

So much for the structure and building of life. Where does evolution come in? Life seems to be composed of three separate sets of "machines"—one for extracting energy from the environment, one for building and maintaining the physical body of the particular life form, and one for maintaining—and then replicating—the information and blueprints not only for the two sets above, but for itself as well. Evolution takes place because of actions by the information system. In fact, it is the very complexity of the information system that allows and sometimes inadvertently prods evolutionary change.

Replication is by far the most difficult process required by life, more so than either building structures or extracting energy from the external environment. DNA and RNA are extremely complex molecules and are necessarily large, even in the simplest of organisms. It now seems that about 200 separate genes are needed for the simplest Earth life. This is compared to about 15,000–25,000 genes in humans, and even more in some other animals and plants. This is far too many genes to put on a single strand of DNA, so life has resorted to multiple strands (chromosomes), each of which has to be replicated.

Highly important to evolution are changes to the genome caused by mutations. This causes a code change, and it can occur either on the chromosome itself during nonreplication times or during replication as a result of any number of replication mistakes. Most such changes are deleterious, causing more harm than good. But they really can change the nature of a gene pool when a change increases fitness of an individual.

Finally, to really bring about variability, sexual reproduction cannot be beaten. It is no wonder that the largest evolutionary rates, and the appearance of so many evolutionary novelties, postdated the evolution of sex.

It is thus the very complexity of life that leads to the mistakes—few enough, but over the long roll of time quite sufficient to continuously reshuffle the deck of genes of any species.

Life seems to have appeared on this planet somewhere between 4.1 and 3.7 billion years ago, somewhere near the end of the Hadean era, or early in the Archean era—or some 0.5 to 0.7 billion years after the Earth originated. Perhaps it is older still, going all the way back to 4.4 billion, the time when liquid water may first have appeared on Earth. However, this is a window of time early in the Earth's history when no fossils were preserved, thus obscuring our understanding of life's earliest incarnation. The oldest fossils that we do find on the planet may be from rocks about 3.6 million years of age, and they look identical to bacteria still on Earth today. (But there is still a debate whether these are indeed fossils of life, or inorganic precipitation of limestones that look like later, layered life). There may have been earlier types of life not now represented on Earth, but our present knowledge suggests that simple oval or spherical bacteria-like forms were the first to fossilize and may have been the shape of the first life on Earth as well. By the time that these appear in the fossil record, we can be sure that evolution was well under way.

COULD THERE BE NON-EARTH LIFE, AND WOULD IT NECESSARILY BE NON-DARWINIAN LIFE?

It now seems reasonable to assert that all known Earth life is Darwinian. Would it be possible for there to be "non-Darwinian" life—life that does not evolve? It is possible to imagine alternative biochemistries of life. Let us take a brief diversion from the themes of this book to see the possibilities. These can be broken down as follows:

1. *Life using different amino acids.* One of the most compelling observations supporting the notion that all life on Earth is descended from a common ancestor is the planetwide use of the same twenty amino acids as the components of encoded proteins. This biochemical uniformity is not obviously demanded by prebiotic chemistry.

2. *Life with chemically different DNA.* An analogous conclusion for terran genetic matter is now possible based on many experiments in synthetic biology. As with alternative amino acids, it appears that DNA molecules using a different "code" not only can work but also can reproduced. For example, an artificial genetic system synthesized in labs at the University of Florida has sustained up to twenty generations of replication (Sismour and Benner 2005). They can even be copied with mutations, where the mutations are themselves replicable. Thus these synthetic genetic molecules are artificial Darwinian chemical systems, according to the research group in Florida headed by biochemist Steven Benner.

3. *Life with a different solvent.* General experience in chemistry suggests that metabolism can operate efficiently only when metabolites are dissolved. Water is an excellent solvent, by many measures. But many compounds are not soluble in water, and indeed there may be habitats elsewhere in the solar system where solvents remaining liquid at either higher or lower temperatures than the 0–100°C range of water would be necessary for any life to exist there. Several of these are shown in figure 1.1, with their temperature ranges.

While the various life forms described above would all have to be classified as "aliens," in one way they all are similar to Earth life: all

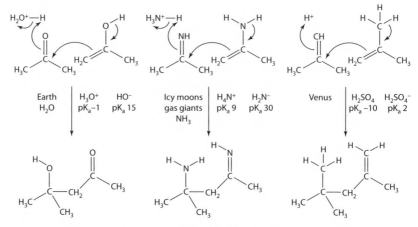

Figure 1.1. Different solvents would favor different, although analogous, chemical reactions to support metabolisms in life residing in different points relative to the star in a solar system. Here are shown three analogous mechanisms for forming carbon-carbon bonds, where the desired reactivity is conferred upon the reacting species by a C=O unit (favored in water), a C=N unit (favored in ammonia), or a C=C unit (favored in strongly acidic solvents such as sulfuric acid). Source: Benner et al. (2004).

should be able to evolve (or there is no chemical reason that they could not). But is there any way there can be nonevolutionary life? In 2003 and 2004 a United States National Academy of Science panel looked at potential chemistry and metabolism of aliens, using the various forms described above as potential candidates. But when they also explored what could be even more alien varieties of what they euphemistically called "Weird," the group concluded that none might be so weird as a potential life form that does not include Darwinian evolution in its makeup. The panel felt that non-Darwinian life is on the other side of the divide between weird (the varieties listed above) and what they called the "truly weird." That line also demarcates the "possible" from the "improbable."

Let us assume that there is non-Darwinian life. Life that does not evolve might be necessarily short-lived or perhaps inhabit environments that are so unchanging as to render the need for evolution moot. Oddly enough, it is probable that the earliest Earth forerunners of life were unable to evolve. There may have been spheres of cell wall with primitive metabolic systems that lacked genomes. They came together, operated in a manner that extracted energy from the environment, perhaps even showed a primitive kind of replication, and then died. Perhaps even primitive genomes would allow replication, and more than a single generation would live. Eventually, however, the lack of evolutionary response would cause death. It is when evolution kicked in that life became life as we know it. And with that property, life altered its effect on the physical world, and then on itself.

It seems likely, then, that most life that can be imagined is characterized by Darwinian evolution. The many varieties listed above certainly suggest that while we may find locales in space where terran life could not survive, we may indeed find exotic kinds of life. Yet, if it is Darwinian, and we have fled our Earth to get away from this trait, it may be that everywhere we went, we would find the same problems. We can assume that any planet with Darwinian life will be hazardous to our health.

2
WHAT IS EVOLUTIONARY "SUCCESS"?

Evolution favors genotypes of high fitness but it does
not generally increase fitness of the species as a whole.
—R. Alexander, *Optima for Animals*, 1996

The Pacific Northwest is moisture-shrouded much of the year;
there are perhaps more different names for rain here than anywhere
else in the world. Along the coastlines and islands fringing this region the rain seems a constant, with rain clouds either hanging above
or coming right down to sea level, immersing life within the mist-
bearing clouds themselves. Here and there, however, a few parcels of
drier country exist, due to fortuitous rain shadows from the many
overlooking mountains. One such place is Sucia Island, a tiny island
almost straddling the U.S.–Canadian border, in the green, cold waters of the Straits of Georgia. At night, the twinkling lights of Van-
couver can be seen to the north, along with the few sparkles of the
Canadian Gulf Islands and American San Juan Islands to the south.

Like many of the place names here, Sucia was named by early
Spanish explorers, who navigated (and named) the Strait of Juan de
Fuca and many of its islands and landmarks while on their quest
to find the elusive Northwest Passage. It sits due northeast of the
Olympic Mountains, and since the prevailing storm track comes
from the southwest, Sucia has a third less rainfall than the nearby
mainland, giving it a different, less vegetated appearance from the
land regions with the excessive (i.e., normal) rainfall of this region.
Perhaps because of this, much of the island shows exposed rock, including high rocky cliffs around its complicated perimeter.

The island is all sedimentary rock, but of two distinct kinds. Most of its many points and embayed walls are a coarse, tan sandstone, and rough quarries shaped like gigantic bites were taken out of the various parts of the island during Seattle's first building boom in the early 1900s, when builders used stone to create their lovely Georgian buildings. The sandstones bear bits of plant fragments, but their most distinctive features are the large, fossilized sand dunes creating arcing cross-bedding now tilted at rakish angles, evidence that these rocks were originally deposited in large sand dunes or shallow, wave-swept sea bottoms.

Sedimentary strata are evidence of old piling on, layer by layer in superposition, with the oldest at the bottom, younger going up. The tan, cross-bedded rocks are the youngest sedimentary beds on the island and have been dated to be about 60 million years in age, or soon after the extinction of the dinosaurs. Old as they are, however, they are underlain by even older rock. Covering the entire southwestern part of the island, these underlying strata are darker in color, a deep olive to dark gray, and are finer grained than their overburden. Their most conspicuous features are fossils—beautifully preserved shells dating back to the later days of the Age of Dinosaurs, but animals then living in the sea, not on land. Their enclosing rocks were deposited on a shallow sea bottom, one obviously rich with life, judging from the amazing abundance of fossils. Most are mollusks, although an occasional crab, shark tooth, or echinoderm is preserved as well, and while clams and snails predominate, the real treasures of Sucia Island are its cephalopod fossils, of two kinds: ammonites and nautiloids. Both have pearly shelled interiors, nacre that glistens in afternoon sun, shooting rainbows of colors off their naturally polished shell walls and chambers.

The ammonites are far, far more common. They were among the most abundant predators in the Mesozoic oceans, judging by their fossil numbers, and where found are usually both diverse and abundant: there were many species of ammonites in any one place, much like there are many kinds of reef fish over any coral reef, with scores of their fossil kind present over any large stratal surface.

But they are not the only fossil cephalopod: rarely, among the cornucopia of ammonites, one can find a representative of the other large lineage of shelled cephalopods, the nautiloids. Far older—the stock, in fact, that begat the ammonites—only a single species is found at Sucia, which is usual for any locality where both groups have been fossilized. So, on Sucia, we have many ammonites and a few nautiloids; and of the ammonites, we have many different species, but only one or two of the ancestral stock.

The difference in number is not the only major difference between these two groups. If one takes a small boat northwest from Sucia, sneaking over the Canadian border like the generations of smugglers who have peopled these islands since the time of the first Spanish and British explorers (getting to the Customs dock will require us to go way out of our way, and we are only jumping to the next island, after all), the next large island is named Saturna. Like Sucia it is made up of the same dark, water-lain strata, similarly packed with fossils. Here too the beautiful ammonites can be found, and the rarer nautiloids as well. But while the strata look the same, the ammonites are mostly different species from those found on Sucia. Only one or two of the fifteen or so different ammonite species common on Sucia are found here. Such a discrepancy can only mean that these two islands are of different age, and sure enough, dating of these two islands shows that the fossils on Sucia are perhaps a million years older than those on Saturna.

We jump back in our boat and head to Saltspring, yet another island in Canadian territory, this one with strata a million years *older* than those on Sucia. Here too the fossils are common, yet here too there is once again a different assemblage of ammonites. But one old friend can be found: the same nautiloid species found on Sucia and Saturna is here on Saltspring as well—again very rare, but seemingly resistant to extinction. In fact, that nautiloid fossil is a species of *Nautilus* and is perhaps even the same species, *Nautilus pompilius*, that is found in small numbers deep in front of the southwest Pacific Ocean tropical coral reef fronts of our own world. And even these ancient members of the genus *Nautilus* are not the first: in the European Jurassic, in rocks 180 million years in age, virtually the same species of *Nautilus* can be found.

Here is an interesting difference. One stock, the ammonites, composed of many hundreds of genera and thousands of species during their time on Earth, evolved quickly, went extinct easily, yet were enormously common. The other, *Nautilus* (which is absolutely typical of the other dozen or so nautiloid genera of the last 200 million years), evolved but a few species, was always rare, but was virtually extinction proof. Now for the difficult question, the subject of this chapter. Which was more successful? What is biological, or evolutionary, success?

And what of we humans? We are but one species, but very abundant. Are we a successful species? Perhaps that will have to await our fate, seeing our ultimate longevity—if we deem longevity a hallmark of success, that is.

THE ELUSIVE CONCEPT OF EVOLUTIONARY SUCCESS

What makes for "successful" life? Just attaining the living state seems daunting enough: there are no more complex assemblages of atoms on Earth than living organisms, and the inorganic process that led to the organic state still defies (and may ever defy) scientific demonstration in the test-tube laboratory. Perhaps we can say that being alive is success. But clearly this is too limiting. We see around us unequal amounts of life, unequal numbers of individuals within individual species. The fossil record such as that starting this chapter yields even more examples of what might be called winners and losers, the abundant and the rare, the long-lived and the short-timers. Defining biological success not only is difficult but may be ultimately self-defeating, for success more often than not is a subjective term. Certainly, since the arguments to come depend on a subjective to objective comparison of success, they are at best anchored in common sense, if nothing else.

With the ignoble caveats of the previous paragraphs, let us return to a primary question: what is success in a biological context? The possibilities are numerous, but those that are clearly primary are far less so. Let us look at these, in no particular order. One way to do this is anthropomorphic: what attributes and/or histories impart success on any human individual?

The success of any human can be measured, or described, in but a few different kinds of metrics. But even these are difficult to classify. I tried to better understand the concept of success, at least as it relates to us humans, by querying a randomly chosen group of well-educated humans. In no particular order, here is the list that they came up with. It surely shares many of the answers that any group of humans might come up with, and it will become apparent by the end of the list that more than a bit of redundancy creeps in:

Money (financial security), a good job, job satisfaction
Good health
Youthfulness
Happiness, contentedness, lack of worry, low stress
How you look—looking good!
Longevity
Spirituality
Quality of life
Fecundity
Emotional support
Lack of pain
Emotional/social circumstances: number of friends
Transcendence
A happy family

Clearly, this group's version of personal success revolved around health and resources. But also amid this list are aspects of success more difficult to quantify but still somehow related to concepts of a successful person: contentment, happiness, and well-being. Yet are these concepts applicable in any way to understanding the concept of success for life, or to any given nonhuman species? Some are, some are not: rating the transcendence and spirituality of a leech, for instance, would be a daunting task.

Thus, with some acknowledgment of the difficulty of such a qualitative and value-dependent task, let us bite the bullet and try to look at a number of possible ways of judging relative success of a particular species. As we shall see, all are to a lesser or greater extent flawed, and perhaps the whole exercise is futile to the point of being nonsensical. The list that seems relevant is listed below in no particular order. There are surely other measures that could be used, but this is the list that comes to my mind, after considerable reflections, as I conduct this work.

1. *Individual (lifespan) longevity.* Using this metric of the average length of time an individual of a given species lives, or dies due to old age rather than early death by accident, disease, or predation. African parrots and some giant tortoises, each lasting in excess of a century, would be among the most successful of all animals; there are some clam species that last four hundred years as well. But there are other animals, such as some species of anemones, that might be functionally immortal if shielded from accidental death; it is the low likelihood of the latter situation, however, that impugns this possibility, as predation and/or fatal environmental perturbation or accident are pervasive. In the plant kingdom, there are individuals that put most animals to shame, such as redwood trees, which have been dated at two thousand years of age, and the bristlecone pines, at up to five thousand years of age.

2. *Species longevity.* The application of radiometric dating methods to fossil-bearing sedimentary rocks during the latter half of the twentieth century provided a wealth of information about the geological longevity of organisms leaving fossil records. One of the surprises of this work was that there was a range of average species longevities—the time between the first appearance of a new species in the fossil record and its disappearance through extinction, rather than some characteristic number for all species. This was interpreted to mean that some species evolve faster than others. The most rapidly evolving were the most useful for biostratigraphy, the science of differentiating the stratigraphic record into successive rock (and hence time) units using successive fossil species. Among these were ammonites, shelled cephalopods related to the extant *Nautilus*; foraminifera, single-celled protozoa with calcareous shells; and, on land, mammal species. At the opposite end of the spectrum are very long-lived species (geologically speaking at least—there does not seem to be a correlation between individual and species longevity), species

referred to as "living fossils." Examples of these are many bivalved mollusks, sharks, and crocodiles. These species last long periods of time because, for whatever reason, they are more "extinction-proof" than most other species. Therefore, perhaps a low extinction rate could be used as a measure of success.

3. *Species "fecundity."* Another unexpected insight from the characterization of average species longevities discussed above was the realization that evolutionary rate or longevity correlated with the number of new species produced over time by any given taxon. Just as any mated pair of humans can produce a range of children, so too can species produce a range of new species. As the fossil record showed, in case after case, that new species mainly formed through processes characterized "punctuational" (after the famous Punctuated Equilibrium model of Niles Eldredge and Stephen Jay Gould), it became clear that species did not smoothly change from one to another, but that one species could give rise to a number of new species, and then exist alongside them. The champion species producers were the same that showed short longevities in time, while long-lived species produced very few new species. For example, we know from the fossil record that over the last 200 million years, species of nautiloid cephalopods averaged durations of over 20 million years each (with some far longer) but produced very few new species. Ammonites, on the other hand, with average species longevity of less than 2 million years, produced many new species. Thus, perhaps success could be measured as the number of new species produced.

4. *Individual abundance.* Another obvious measure could be the number, or biomass, of a particular species. While there are many rare species, (and many more of such species every year during this dominion of humankind), some species are exceedingly common, such as common weeds like dandelions, English sparrows, fruit flies, and many kinds of microbes. Using this measure, it may be that some varieties of viruses or bacteria are the most common organisms on the planet.

5. *Percentage of the planet's biomass.* This measure is somewhat related to the preceding. Perhaps not simply the number of individuals can measure success, but the proportion of the total biomass on

Earth an individual species makes up. The champions of this category would clearly be microbes, perhaps the common cyanobacteria that photosynthesize in oceanic or fresh water, or perhaps the microbes inhabiting the Deep Microbial Biosphere, bacteria and archeans that use hydrogen from rocks for energy.

6. *Species that co-opt other species for their betterment.* Humans have to be considered successful by the fact that we have manipulated so many other species (or wiped them out). Termites and ants also co-opt other species, and have increased in numbers because of this. We humans (and termites and ants as well) also make up a significant terrestrial biomass (although the human biomass is probably dwarfed by the common Norwegian rat and common cockroach, among many other "weeds"). This category cannot be judged simply as a "brain size" metric. Whales, for instance, with very large brains, exist in very low numbers and thus do not pass any other "success" test, while ants have almost no brain at all.

7. *Wide geographic range.* Any species begins at one geographic locality, usually as a peripheral isolate of an existing species. From there, some species move to ever larger geographic ranges. Some become worldwide in extent, either on land or in the sea. This is a function both of geographic longevity and dispersal ability. Perhaps the degree of range can be used as a measure of success.

8. *Surviving mass extinctions.* The many mass extinctions of the geological past have proven to be filters for biodiversity. They are not simply some kind of background extinction rate (during non–mass extinction times there are always some number of extinctions taking place through natural processes) writ large. Some species have gotten through individual mass extinctions, only to be taken down in a successive event.

9. *The ability to move to other planets as the Earth becomes uninhabitable, or to keep the Earth habitable beyond its natural lifetime.* As we will see, the Earth has a finite amount of time as a place that can support life. Perhaps the ultimate success is a species capable of moving to a new habitat as this one loses it habitability, or to slow or stop the loss of habitability. This would certainly increase that species' temporal longevity.

A MODEL FOR A MAXIMALLY SUCCESSFUL SPECIES

By using the various factors above, it is possible to propose the characteristics of an optimally ideal or "successful" species, at least based on this list. The ideal individual would be immortal and would belong to a species that is extinction-proof, is widespread, has the highest relative and absolute biomass on the planet, and has the capability to either terraform our planet as it becomes "unterran" or migrate to a new planet. Terran life fails on all counts with one exception. Way back in our species' adolescence, some 125,000 years ago, who could have predicted our species' eventual success as we emerged in the evolutionary transition from archaic to modern *Homo sapiens sapiens*? And perhaps even more so, beginning some 30,000–35,000 years ago, when a subtle but important mutation resulting in a more practical kind of tool-using intelligence occurred in a small group of our species, a group that ultimately went on to replace those humans without it? We were not pretty; we were not athletes of the dangerous African plains, capable of rapid tree climbing, or speedy enough to escape an onrushing predator. We could not jump out of harm's way, or fly, or even swim very fast. We were essentially cat food until the invention of Clovis technology about 10,000 years ago. All we had going for us were our brains. But those brains would more than get us to the Earth species reunion of all still surviving species, these thousands of years later. Not only are we the major success, we are the only hope for life to save itself from itself.

A MODEL FOR A SUCCESSFUL BIOSPHERE

In the same vein, what would constitute a "successful" biosphere? It appears that the properties of populations and ecosystems are different from the properties of individuals. Altruism, for instance, is known in biology but can be selected for only at the level of population, not individual. So too with social insects, composed of a few breeders and many nonbreeding workers.

22

The choices are somewhat similar to those for the species, but there are also major differences. Here are some potential attributes.

Diversity of species on Earth
Biomass on Earth
Stability of the species assemblage on Earth
Minimal risk of the end of life on the planet through some means

why?

obviously

There are two stark choices that are relevant to the theme of this book. We live today in a very diverse world. Some models, discussed in a later chapter, suggest that there are more species on Earth today than anytime in the past. Personally, I suspect this not to be true. The well-known extinction of megamammals over the past 50,000 years through some combination of overkill and climate change may be just the tip of the iceberg, no pun intended. I would surmise that maximal diversity occurred in the Eocene, with a global jungle and nearly global tropics. But models by a group at Potsdam indicate maximal biomass nearly a billion years ago, with a current downward trend. Yet that would mean that biomass and diversity are decoupled. If so, there is an interesting implication. Could it be that the production of ever more species actually reduces biomass on a planet? Perhaps we have traded a world of a few, long-lived species for one with many, "more successful" species, in that they produce many daughter species of themselves but at the same time have a very high extinction rate. This would be a very Medean relationship, which will be defined in the next chapter.

or not

Speculation

evidence?

Cause-effect?

how so?

Why would decoupled be Medean?

3

TWO HYPOTHESES ABOUT THE NATURE OF LIFE ON EARTH

The first impulse [of Gaian Theory supporters] is to interpret the history of the Earth as an epic tale in which the organisms play heroic starring roles.

—James Kirchner

In late August 2007 the Northern Hemisphere was witness to a spectacular lunar eclipse. It is rare that so large and so populated an area finds itself under the path of totality, but this one did. Near its West Coast totality (the eclipse reached totality in the very early morning hours even on the West Coast of North America), an actor and retired lawyer named Paul Taylor set fire to a four-story-tall wooden man standing serenely on a Nevadan desert at a place called Black Rock. This large effigy was scheduled to burn some days later, at the culmination of the Burning Man Festival, a cult, pantheistic "festival" that, over the years since its inception some two decades ago, has grown to a week of New Age worship (and aging hippie hijinks). The grand conclusion is the huge bonfire, visible across a wide swath of the Black Rock playa.

Unfortunately for Taylor, the organizers of Burning Man were not amused. The cops were called, and Taylor was led off in handcuffs, charged with felony arson. There was a palpable sense of irony in the heavy charge for burning something destined to be burned anyway, but timing is everything, I suppose.

Afterwards the blogs buzzed about this anarchist event, especially after the arsonist explained that his torching was in protest against the commercialism of Burning Man, which had started out on a Bay

Area beach before moving to Nevada. In its early days it was a kind of West Coast Woodstock, but where that iconic festival was about music, Burning Man was meant to be cerebral, and a continuing thread from year to year was the worship of Gaia, the Greek name for the goddess of the Earth. Thus when the blogs spilled over with people's varied reactions to the early burning, more than once, the goddess Gaia was implored to forgive the intemperate arsonist.

As viewed by the Burning Man followers on the Nevada desert, Gaia is a sweet, forgiving mother—perhaps the Earth itself, in reality quite alive. But others have a bit more of a negative, or alternative, view of her. For instance, on a website called Gaia Gone Wild (gaiagonewild.com), Gaia's various rather mean-spirited retorts to human existence are featured as short video clips, including the massive 2007 wildfires that blackened Greece, as well as sundry other tornadoes, hurricanes, and even dust devils. Another such exercise in characterizing Gaia as a vengeful being is a 2006 book, *The Revenge of Gaia*, which was written by the English atmospheric scientist, Sir James Lovelock. It seems as if Gaia is lately taking a rather annoyed notice of us humans. As far back as 1996, the official website of the Burning Man Festival featured Gaia as "The Cruel Mistress." Those seeking forgiveness and atonement were invited to enter a chamber where they would witness "the great mystery of Gaia," where their sins would be purified and they could be "Re-Virginated" (according to the website, at least).

Somehow, a vision of the goddess Gaia covered with chains and black leather and wielding a whip is not exactly a maternal image— particularly if we assume that all of the species on Earth are her "children." However, if we think about it, this image just may be a more accurate one than the New Agers and many current scientists would suspect.

If one digs through the New Age literature in search of the origins of Gaia as a philosophical underpinning, several notable things emerge. First, a number of very vocal Christians have weighed in on Burning Man (in general) and worshiping Gaia (in particular) as the outgrowth of Satan's work; apparently, this festival, and goddess worship, were arranged by evil incarnate to further bring down our species. Second, and more notably, however, there is an intellectual thread running through the literature, one that ties the process of evolution to Gaia, as her major tool for affecting life.

There are many religions on Earth, and Gaia worship in some form or another is surely ancient. It arose as a celebration—and fear—of nature. Its current guise is one that seems to differ broadly from other religions in the sense that it spiritualizes the observation of nature, and the relationship between life and the planet. Inspiration for the Gaia hypotheses, or the idea of thinking of the Earth as an integrated system, comes from ancient ideas of what our relationship with the Earth is and should be; but nowadays, it seems that the flow of ideas is not going from "spiritual connection between life and the planet" to "Gaia hypothesis/Earth system science," but rather from "Gaia hypothesis" to "New Age Gaia-worship." The recent ideas and discoveries of scientists are being co-opted back into a New Age form of Gaia-worship. But just how was a scientific hypothesis so quickly co-opted by a new form of an old religion? Perhaps at the heart of both science and neo-Gaia worship there is a shared, deep-rooted, and intuitive interest in better understanding the true nature of life and its relationship to the only home we know, the Earth. Of course, we'll be concentrating here on what science can tell us about life's relationship to the Earth, rather than any sort of rekindled sense of mystical fascination in this puzzle. And to describe what science has found out in answer to the mystery of the nature of Earth life, we need first to go back to what we might call the basic "characteristics" of Earth life. Then we will be able to go on to discuss Gaia, Medea, and how we can test both of these ideas for validity.

EVOLUTION AND ORGANISMAL BEHAVIOR

The inclusion of evolution as a required element of life (at least of life on Earth) must lead to certain behaviors of that life. Most importantly these include competition even within a species, sometimes cannibalism, as well the behavior of breeding to a resource-defined "carrying capacity"; be it beetles in a jar or humans on our planet, any given species of life seems to breed not only up to the point where all resources are spoken for, but beyond that point, so that there are more individuals than resources can sustain. "Carrying capacity" is a formal concept in ecology defining that limit.

Competition is thus an inherent attribute of Darwinian life. But competition works at the level of the individual. Are there larger scales, so-called macroevolutionary" aspects of Darwinian life, that accrue from competition or other unnamed aspects of the individual?

The field of paleobiology from the 1970s to this time has been scientifically and intellectually driven by studies of macroevolution: does macroevolution exist, and if so, what does it do? Led by such stalwarts as Stephen Jay Gould, Dave Raup, Steve Stanley, Jack Sepkoski, and David Jablonski, a keen appreciation not only of the existence of macroevolution but of its processes is fairly well known at this time. We know the way that new species form, which species are more susceptible to extinction in normal times (as opposed to times of infrequent but devastating mass extinctions), and, most important, whether some species produce more "daughter species" than others, and how the longevity of species is related (if at all) to the rates of new species formation.

In the 1970s another and quite distinct property of life over and above these micro- and macroevolutionary aspects was hypothesized: a property emerging not from individuals, not from species, but from the aggregate of life on Earth at any given moment in geological time. It has been proposed that life improves the planet for itself, and for future life. This was called the "Gaia hypothesis," which later was modified into two distinct kinds of "Gaia," which may be treated as separate and distinct hypotheses. Thus both must be tested in a scientific manner.

GAIA HYPOTHESIS

From classical times it has been tempting to analogize our planet (or any habitable planet, according to some), as itself some sort of living thing. But the formalization of this idea came from the gifted British scientist James Lovelock (mentioned above for one of his many books with Gaia in the title), who, from the 1970s to the present, has eloquently argued this view from a scientific perspective as well as in a form more accessible to nonscientists. One of his quotes gives an idea of the scope of the Gaia hypothesis, or Gaia Theory, as Lovelock and his equally gifted collaborator Lynn Margulis now call it: "Specifically, the Gaia Theory says that the temperature, oxidation, state, acidity, and certain aspects of the rocks and waters are kept constant, and that this homeostasis is maintained by active feedback processes operated automatically and unconsciously by the biota." Because there exist various distinct varieties of Gaia hypotheses, from here on I will refer to Gaia theory as the sum of all its parts, and this is meant to be distinct from Gaia Theory, "with a capital T," as it was once described by Lynn Margulis.

The idea was simple and elegant, and quickly Gaia theory attracted many adherents, both scientists and nonscientists alike. Some researchers saw in Gaia theory a new way of thinking about the cycles of life's organic components and elements. Some followed Lovelock's lead in searching for scientific support for the idea that the biomass of Earth life self-regulates the conditions on the planet to make its physical environment (in particular temperature and chemistry of the atmosphere) more hospitable to the species that constitute its life. An even stronger claim that also appeared early in the history of the Gaia hypothesis is that all life-forms are part of a single planetary being, called Gaia, or modeled as such ("geophysiological Gaia"). Lovelock does not now subscribe to this latter view, although he certainly seemed to see the Earth as "alive," based on the following quote: "The entire range of living matter on Earth from whales to viruses and from oaks to algae could be regarded as constituting a single living entity capable of maintaining the Earth's atmosphere to suit its overall needs and endowed with faculties and powers far beyond those of its constituent parts."

Another quote in this vein is from Lovelock's first book on Gaia (published in 1979): "the quest for Gaia is an attempt to find the largest living creature on Earth."

While a view not subscribed to by any current scientist to my knowledge, this early interpretation galvanized some of the more passionate supporters. The "planet is alive" interpretation is the "strongest" form of Gaia. (The use of "weak" to "strong" aspects of Gaia theory, now commonplace when discussing various components

Really?

How many asked?

of the concepts, was first made by atmospheric scientist James Kirchner, a critic of Gaia theory).

Over time, Lovelock came to amend his original definition of the Gaia hypothesis, abandoning its more extreme forms. Yet while Lovelock may have changed his statements about Gaia, many of those who had heard about Gaia through news reports, which often emphasized the more extreme—and thus newsworthy—implications of Gaia theory, took no notice of the scientists' caveats. The power and simplicity of a living planet ruled by a benevolent goddess filled a void among many of the baby boomers who, during the 1960s and 1970s, were turning away in large numbers from what they saw as the conservative, restrictive, traditional religions. Here was a natural successor to those traditional religions, and one that could comfortably coexist with both science and the emerging environmental ethic. No miracles were required to believe in Gaia, since her elastic and multiple definitions gave her a really big tent.

And what of the scientists? Lovelock was not the only scientist involved in Gaia theory, and definitions of the central hypothesis evolved through time. There now appear to be three (or more, depending on whom one talks to) distinct and valid categories of the Gaia hypothesis, all different from one another. (I am excluding the "planet is alive" hypothesis from further discussion; it is firmly rejected.)

A further prediction from Gaia literature is that organisms not only combine to *maintain* living conditions, but also in fact will ultimately *extend* the lifespan of the Earth. We will return to this aspect of Gaia theory in a subsequent chapter. First, let us look at scientific implications of the Gaian literature.

The first two are sometimes referred to as "Healing Gaia." They thus will be treated as separate hypotheses in the remainder of this book. They are, in order of appearance, "Optimizing Gaia" and "Self-regulating Gaia."

HYPOTHESIS 1. Optimizing Gaia
This early interpretation remains one of the "strongest" versions of Gaia theory. It implies that there is actual control of environ-

mental conditions, including purely physical aspects of the biosphere, such as temperature, oceanic pH, and even atmospheric gas composition. Here are four quotes from the 1970s and 1980s by James Lovelock and two coauthors, arranged in chronological order from oldest to youngest, that epitomize "Optimizing Gaia." The first was:

> Those species of organisms that retain or alter conditions optimizing their fitness (i.e., proportion of offspring left to the subsequent generation) leave more of the same. In this way conditions are retained or altered to their benefit. (Lovelock and Margulis 1974, p. 99)

Then, years later:

> The Gaia hypothesis ... postulates that the climate and chemical composition of the Earth's surface are kept in homeostasis at an optimum by and for the biosphere. (Lovelock and Watson 1982, p. 795)

Four years later this was written:

> Life and the environment evolve together as a single system so that not only does the species that leaves the most progeny tend to inherit the environment but also the environment that favors the most progeny is itself sustained. (Lovelock 1987, p. 13)

And finally:

> The system of organisms and their planet, Gaia for short, must be able to regulate its climate and chemical state. . . . the greater part of our own environment on earth is always perfect and comfortable for life. . . . Through Gaia theory, I see the Earth and the life it bears as a system, a system that has the capacity to regulate the temperature and composition of the Earth's surface and keep it comfortable for living organisms. (Lovelock 1988, pp. 7–8, 30)

Another way of interpreting this hypothesis is that the biomass alters conditions on Earth to increase the "hospitality" of the planet. This was scientifically termed "full homeostasis" (a word very difficult to define, I find). Another version of this comes from the thoughtful work of Tim Lenton, who has proposed that life has managed to increase the resiliency of the entire biosphere in response to external perturbations or shocks.

Scientists belonging to a new field called Earth System Science, a direct offshoot of Gaia theory (and now a well-funded, rigorous area of science) followed Lovelock's lead in searching for scientific support for the idea that the biomass of Earth life self-regulates the conditions on the planet to make its physical environment (in particular temperature and chemistry of the atmosphere) more hospitable to the species that constitute its life. (The many recent texts on Earth system science, however, do not entail a commitment to Optimizing Gaia.) However, there is a degree of obvious circularity in the argument that the Earth is ideally suited for life, because the life present on Earth has had a long time to adapt to the very conditions on Earth. As James Kirchner noted in 1992 (p. 399),

> The life forms that we observe today are descended from a very select subset of evolutionary lineages, namely those for which Earth's conditions have been favorable. The other lineages, for which Earth's conditions are hostile, have either gone extinct or are found in refugia (such as anaerobic sediments), which protect them from the conditions that prevail elsewhere. As Holland (1984, p. 539) has put it, "We live on an Earth that is the best of all possible worlds only for those who are well adapted to its current state."

During the 1990s Lovelock backed away from Optimizing Gaia. Even he found it too extreme and unscientific, apparently.

HYPOTHESIS 2. Self-regulating (or Homeostatic) Gaia;
Negative Feedback Gaia
The more recent addition to Gaia theory can be called "Self-regulating Gaia," which supposes that feedback systems allow the continuation of life on Earth by keeping life-constraining factors such as temperature, and more recently even atmospheric

oxygen and carbon dioxide levels (the latter directly affecting planetary temperature), within ranges that allow life. This can be rephrased as a question: Do biological feedbacks stabilize, or destabilize, the global environment? The Gaia literature indicates that most or all of the feedbacks are *negative*; thus if the planet's temperature rises to dangerous levels, actions of organisms help bring it down. Thus feedbacks should be negative, and the potential existence of positive feedbacks has been suggested as a way of testing this version of Gaia theory.

This hypothesis has itself been interpreted in two ways. The first is that life tends to make the environment stable, allowing all life to flourish. The second is that life tends to make the environment stable *in order* for all life to flourish. This is a subtle but important change of wording, for the latter implies intent.

HYPOTHESIS 3. Coevolutionary Gaia

James Kirchner presented this hypothesis in 1989 at the first organized scientific conference dedicated solely to Gaia theory; it simply advocates that the biota and environment have evolved in a coupled way. This statement is the "weakest" of the hypotheses, as it was already viewed as true.

HYPOTHESIS 4. Progressive, Deterministic Gaia

This is essentially a stronger version of so-called Coevolutionary Gaia. It is explained in David Schwartzman's important book, *Life, Temperature, and the Earth*, and even more recently he has described it thusly [p. 207]:

> The evolution of the terrestrial biosphere is quasi-deterministic, i.e., the general pattern of the tightly coupled evolution of biota and climate was very probable and self-selected from a relatively small number of possible histories at the macro scale, given the same initial conditions. . . . Major events in biotic evolution were likely forced by environmental physics and chemistry, including photosynthesis, as well as the merging of complementary metabolisms that resulted in new types of cells (such as eukaryotes) and multicellularity.

In other words, once life evolved, there came into existence only a few possible pathways for how life and its systems would evolve further. Nothing special here, simply a small number of nutrient and element cycles that themselves affected later life.

Other Gaian Views

Garbage Can Gaia

There are many other interpretations of Gaia—enough to fill volumes of books. One of my favorites, however, comes from Tyler Volk, who has proposed "Wasteworld Gaia" (hence my own, more colorful appellation starting this section). Though he is not proposing as much of a hypothesis as those above, Volk suggests that the atmosphere, the very clue that led Lovelock to the whole Gaia concept in the first place (because of its being in chemical disequilibrium), is nothing but one giant waste dump. In this not-so-charming view, life produces waste material—prodigious quantities of waste, which build up and affect the environment and, of course, the organisms living in the environment. This was certainly the case when life began pouring oxygen into the atmosphere at a time when oxygen was poisonous to most life on the planet. Volk's "Wasteworld" has waste building up to the point that it becomes intolerable to certain kinds of life. But then along comes some new kind of life that uses the waste in some novel way. Volk even describes Gaia as nothing but the collective byproducts of life itself. He wondered as well if the accumulation of these byproducts could in fact cause the increase in resiliency of the biosphere mentioned by Tim Lenton.

Evolving Gaia

Another view proposed by Lenton is that life has not survived for over 3.8 billion years purely by chance, but rather because of the Earth system's regulatory mechanisms. Furthermore, he believes that the Earth system with life present is more resistant and resilient to many (but not all) perturbations, and that the

Earth is predicted to remain habitable for longer with life present than it would without. I will reexamine this particular point in detail in a later chapter because I find myself in extreme disagreement with almost all of Lenton's conclusions. There was and will always be a change of the physical Earth because of the presence of life in myriad ways, and changes as well to many Earth "systems," such as the carbon and hydrological cycle, to name but two. But Lenton's very Gaian view is that as the biota evolves, its ability to actually regulate the systems evolves as well, and that these regulatory properties not only accumulate but strengthen as the biota evolves. Thus the intersection of life and the environment creates various systems, and over time these systems evolve in such a way as to make life more resilient in the face of environmental shocks and perturbations (such as more sunlight, less tectonic activity, and the occasional and major asteroid impact). A corollary is that the surface environment of the Earth changes less, and recovers more quickly in response to perturbation, with life on it than it would if life were not present at all.

Seems empirically true

HYPOTHESIS 5. The Medea Hypothesis

The Gaia hypotheses are evidently both powerful and influential. But are they testable? There is now a rather long list of papers questioning the various Gaia hypotheses, with the most important the thoughtful reviews by Nobel Laureate Paul Crutzen and Berkeley atmospheric scientist James Kirchner. I have borrowed heavily from both of these eminent scientists' works in many of the arguments to follow. Most crucially, however, both Crutzen and Kirchner point to the same problem—neither Optimizing Gaia nor Regulating (homeostatic) Gaia is readily testable by the scientific method. And Coevolutionary Gaia is virtually no hypothesis at all. *No!*

Not testable – but true of many/all N=1 systems (ecosystem, cosmology etc)i What about p 32 ⚹

Any scientific hypothesis must be both testable and predictive, and while our overall understanding—of how various elemental cycles (such as the carbon, oxygen, and sulfur cycles) help to maintain life on Earth—has improved as a result of developing and debating the above

34

hypotheses, testing either the Optimizing or the Self-regulating Gaia hypothesis has proven difficult, in no small part because of the rather nebulous definitions of the hypotheses themselves. Thus many critics of these hypotheses wonder if they are science at all.

My goal is to propose a new hypothesis that can explain a variety of events and characteristics of Earth life. Thus here I propose what I call the Medea hypothesis, which can be formalized as follows. Habitability of the Earth has been affected by the presence of life, but the overall effect of life has been and will be to reduce the longevity of the Earth as a habitable planet. Life itself, because it is inherently Darwinian, is biocidal, suicidal, and creates a series of positive feedbacks to Earth systems (such as global temperature and atmospheric carbon dioxide and methane content) that harm later generations. Thus it is life that will cause the end of itself, on this or any planet inhabited by Darwinian life, through perturbation and changes of either temperature, atmospheric gas composition, or elemental cycles to values inimical to life.

ON THE NATURE OF MEDEAN LIFE

What are the characteristics of this kind of life? Some are self evident: all of the Earth life variety, anyway, has a finite life span. All has a series of environmental tolerances that are a subset of conditions found on Earth. There are other such obvious aspects. But what of the less obvious, but nevertheless fundamental, properties of life that play important roles in regulating the life of the biosphere. Are these "Medean"?

1. All species increase in population not only to the carrying capacity as defined by some or a number of limiting factors, but to levels beyond that capacity, thus causing a death rate higher than would otherwise have been dictated by limiting resources.

There are any number of examples of this. Put any sort of breeding pair of insects in a closed jar with a finite amount of food. The bugs quickly multiply, the food disappearing more and more quickly until it is gone, and the bugs die off from starvation, usually with some last phase of cannibalism preceding the complete extinction.

Now start the experiment anew, but this time put a constant amount of food in the bottle over time. The number of bugs rapidly increases and then stabilizes at some number dictated by the food level; this was a classic experiment in ecology and is found in every ecology textbook. But the point here is that the number of bugs is still too high for the amount of food. In fact, there are always too many bugs for the amount of food, and some number are always dying of starvation or being killed by other bugs as they fight for food. The population does not reach a stable number in peaceful coexistence. There are always too many bugs causing intense intraspecific competition. This is really at the heart of Darwinian evolution as elucidated by Darwin. It is in his principle of variation: there are always more young than can be supported by resources, and "survival of the fittest" ensues.

This "property" of Darwinian life is universal across the taxonomic spectrum. Humans are no exception. The many anthropological fairy tales of human population as self-regulating to match resources have been exploded case by case, and Jared Diamond's recent book *Collapse* provides innumerable examples.

2. Life is self-poisoning in closed systems. The byproduct of species metabolism is usually toxic unless dispersed away. Animals produce carbon dioxide and liquid and solid waste. In closed spaces this material can build up to levels lethal either through direct poisoning or by allowing other kinds of organisms living at low levels (such as the microbes living in animal guts and carried along with fecal wastes) to bloom into populations that also produce toxins from their own metabolisms.

3. In ecosystems with more than a single species there will be competition for resources, ultimately leading to extinction or emigration of some of the original species.

4. Life produces a variety of feedbacks in Earth systems. The majority are positive, however.

Can these properties shed light on the nature not of individual species, but of the biosphere? Under Gaian sense, these rather selfish biocidal and suicidal tendencies of populations somehow change at the scale of the biosphere. While the beetles overbreed and kill themselves off, the entire world of beetles and everything else somehow manages to improve conditions for life, so that biomass and diversity increases through time. Individuals and populations are seen as bad, but together they all do "good." I would suggest that individuals are neutral, but life as an aggregate is negative to itself.

Let us make another set of predictions, one that we might call "Medean."

1. Diversity and biomass can be independent and decoupled.

2. The history of biomass (not diversity!) should show a series of steps—from the first formation of life, to various kinds of life harnessing ever more energy through better metabolisms. After some period of time, however, each group of organisms that utilizes the new kind of "energy regime" (actually a new kind of metabolism, a way of getting energy during everyday life) will show a slow decay, to biomass levels lower than those following the initial diversification of the organisms. In other words, we should see a mass extinction of the preponderance of organisms of the preceding metabolism, through mass extinction of organisms of the preceding regime.

3. In all but regions of low environmental disturbance, ecosystems will eventually move toward lower species diversity as the competitive forms drive other species into extinction and as some species move to dominance.

These characteristics of life are pretty self-evident. In his review of the Gaia hypothesis, Kirchner (2002, p. 403) explicitly noted the rather destructive inherent nature of life:

All organisms must consume resources, and by doing so they deplete their local environments of those resources. Likewise, all organisms must eliminate wastes, and by doing so they pollute their environments. Traits that enable organisms to better consume resources or eliminate wastes will benefit the individual, and thus will be favored by natural selection, even though they also degrade the environment. Examples of such traits abound. Trees are highly evolved to catch sunlight, and thus shade their neighbors. Plants in arid zones are highly evolved to intercept moisture before it reaches their competitors. Some tree species (such as eucalyptus and black walnut) even conduct a form of chemical warfare against potential competitors, by dropping leaves or fruits that make the surrounding soils toxic for other species.

TESTING THE HYPOTHESES

Here are three generalities that appear to falsify the various Gaia hypotheses.

1. Gaia theory predicts that biological feedbacks should regulate the Earth's climate over the long term, but peaks in paleotemperature correspond to peaks in paleo-CO_2 in records stretching back to the Permian. This is because carbon dioxide is a potent "greenhouse gas"—it holds heat in. When CO_2 rises in the atmosphere, much evidence shows that temperature rises (or in the past, where most evidence comes from, that temperature rose). Thus if CO_2 is biologically regulated as part of a global thermostat, that thermostat has been hooked up backwards for at least the past 300 million years.

2. Gaia theory predicts that organisms alter their environment to their own benefit, but throughout most of the surface ocean (comprising more than half of the globe), nutrient depletion by plankton has almost created a biological desert—a place with little life in it. This seems counterintuitive—we think of the ocean as a place thriving with life. But this is not the case over most of the ocean's surface, especially those areas far from shore,

3. Where organisms enhance their environment for themselves, they create positive, rather than the predicted negative (healing Gaia), feedback; thus Gaia theory's two central principles—that organisms stabilize their environment, and that organisms alter their environment in ways that benefit them—are mutually inconsistent with one another.

Specific tests of these two hypotheses will be the subject of the next two chapters. Before that, however, the actual systems that affect life, its abundance, and its longevity need to be briefly described. They are part of Earth system science, mentioned above.

Earth System Science and Tests of Gaia

There are indeed highly complex "life support systems" that produce planetary conditions, such as atmospheric composition and pressure, planetary temperature, and even geomorphic surface features, that are quite different from those of a lifeless planet. The workings of these life support systems and their changes through time are measurable, the effects testable. The most important tests, however, were made by the Earth itself, in a series of Earth history episodes. These many episodes (chosen as the most important from a large number) will be the tests themselves. Each of these times marks a short period when the amount of life on the planet dropped precipitously, in each case the cause was planetary poisoning brought about by some variety (or varieties) of life itself, and in each case it can be demonstrated that recovery was a long process.

To support (and presumably test) Gaia theory, Lovelock and Margulis challenged scientists to search for the current factors that keep life alive on our planet—the carbon, sulfur, phosphorus, iron, oxygen, and hydrological cycles (among others) that nurture life—and to search as well for clues in Earth history supporting their view. Many scientists did answer this call. There is now a whole branch of science, Earth System Science, that either directly or indirectly gathers ever more information about the chemical cycles necessary to succor life. The formation of this exciting and valuable field is a direct result of the formation of the various Gaia hypotheses, and thus a great debt is owed to the originators of the hypotheses.

To say that Gaia theory was influential is an understatement. Very few hypotheses or even theories move out of the realm of science into the other philosophies of we humans. But Gaia has, becoming the lynchpin of the New Age movement, as well as providing a moral imperative: that we must treat this living planet not as hunter/gatherers but as agriculturalists and preservationists. Gaia thus becomes a foundation of modern environmentalism, which takes the stand that human activity has caused deleterious changes in the Gaian workings, and that the original, pristine cycles must be restored. Many would have us believe that our planet should be put on the endangered species list, and that if only humans would somehow disappear, things would return to their natural order.

The science behind Gaia is thus both about current processes and about history—not just what happened, but how the planet and its life support systems evolved through time. It also sets the stage for understanding other planets. When the Earth is viewed not as a unique entity, but as one of surely many habitable planets, an entirely new kind of understanding arrives. The concept of "habitable planets" is based on planetary nurture, with life being the ultimate and hoped for result of planetary formation and change.

So what are the specific Earth systems—and what is a system at all? A system can be defined as a group of components that interact. They are thus interrelated parts that function as a complex whole. In a vertebrate animal (and many invertebrates as well), the interacting systems include the circulatory, respiratory, endocrine, nervous and sensory, lymphatic, excretory, digestive, and reproductive systems, among others.

Organisms need physical material and energy to grow. They need the matter necessary to build cell walls and organelles, nucleic acids and polymers, the entire physical superstructure that is life. Organisms are open systems: all need the addition of material during life for life and growth. Humans—and almost all other organisms—do not last very long without a constant intake of new material. In this, life and the planet are very different. Planet Earth is a closed system with respect to material—essentially we do not receive new material from outer space, but continuously recycle what is already present—but an open system with respect to energy, whereas all organisms are "open" systems with respect to both. Since no new material needed for building living things is being added to the planet, the building blocks of life have to be recycled. This is accomplished by a series of element and compound circulation systems.

For life, the most important of these fluxes are the movement and transformation of the elements carbon, nitrogen, sulfur, phosphorus, and various trace elements. Each of these elements is critical to the existence of life on this planet. These and other elements move in and out of the atmosphere, hydrosphere, and the solid Earth. Because movement and transformation of matter require energy, Earth system science also examines the energetic underpinnings of the various systems, which largely come from two sources: the Sun, and heat generated from the breakdown of radioactive material beneath the Earth's surface.

Each of these systems has changed through time and will continue to do so. The presence of life on the planet and the ability of life to evolve and increase in complexity through time have caused each of the nonbiotic Earth systems to be modified, and then caused feedbacks affecting life. These couplings linking the organic and inorganic components of the Earth have evolved in tandem over time as the Earth has aged, and as life has radically transformed itself into increasing diversity and complexity. The study of the Earth has yielded accurate information about how these interactions have occurred in the past, which allows us to make predictions about how these systems will change in the future.

Gaia theory proposes that the many Earth systems change in ways that make life itself more "successful"—not any individual species, but the biota of the whole, the biota being the sum total of all organisms living in, on, and above the Earth. To test this, we first need to look in more detail at the systems themselves—how they work, and how they have changed through time. Here are the systems identified by Earth system science.

Solar Lighting and Heating

Life requires energy. Be it a plant directly harvesting energy from the Sun or an animal ingesting some part of that plant and then acquiring solar energy second (or third or fourth) hand, the Sun is the ultimate source of almost all of life's energy on Earth.

Big, bright, and beautiful, the Sun is our source of warmth and light in a universe of cold and dark. Not only does the Sun's light energy power photosynthesis, but also its gravity holds us in orbit, and its heat keeps us from freezing. It powers the wind, drives the waves, and makes clouds that provide an ocean-covered planet with a nearly endless supply of fresh water—all of this from just an incandescent ball of gas. But what a body it is, radiating 60 million watts of power from every square meter of its surface. Even at our distance of 150 million kilometers, the light of the midday summer Sun still carries over a kilowatt of power to each square meter of Earth it illuminates. The energy is immense: even the smallest towns receive over a billion watts of free power from our brilliant star.

Unfortunately, like many other good things, the Sun also has a dark side. The star that we orbit is a time bomb, and each tick moves us toward a future of drastic change. After nourishing life for billions of years, the Sun will evolve and cause many of what can be called the "ends of the Earth." It will be *the* major factor that drives the Earth to its ultimate fate, but as we shall see, that fate has also been preordained in no small way by the role of organisms. It is truly ironic that the Sun, a body that has played such a positive role in the Earth's history, is also one of the villains, responsible for our planet's ultimate demise.

The Sun is a powerful nuclear reactor, but how stable is it? We have no direct long-term record of the Sun's output, and our only direct insight comes from observation of similar stars. Stars that have a similar mass and age as the Sun also intrinsically have nearly the same brightness as the Sun. This suggests that the brightness of these Sun-like stars does not vary greatly. It is expected, however, that long-term changes must occur. We are certain that, very slowly, the Sun gets brighter. Stars are essentially great energy-generating engines. They work fabulously for billions of years, but near the end of their lives they begin to run out of fuel and undergo extraordinary and complex changes. Unlike most other engines, they do not fade as they age but become ever more powerful and energetic.

The Sun becomes brighter because the number of atoms in its center is decreasing. I imagine a balloon enclosing the core of the Sun. The pressure on the inside has to exactly support the weight of the overlying mass of the Sun. We know that the size of the Sun does not change over long periods of time, so the pressure in its center must remain reasonably constant. The pressure is produced by the cumulative impacts of vast numbers of particles. As each atom bounces off the surface of the imaginary balloon, it imparts a small outward force. The total pressure is the net effect of all of the particles in the balloon. Using what is known as the "Gas Law," the pressure in the balloon depends on just two things: the number of particles in the balloon and the temperature of the gas. As we will see, the number of particles is constantly decreasing and so the temperature must constantly rise if the pressure is to remain constant.

As the Sun evolves, the number of particles in the balloon de-creases. The chain of nuclear reactions effectively takes four protons and turns them into one helium nucleus. If all the hydrogen were converted to helium, the number of particles left would be only one fourth of the number that the Sun started with. As the number of particles gradually decreases, the temperature of the Sun's core rises. As the temperature rises, hydrogen travels at higher speed, collisions are more energetic, and the production of helium and the total amount of energy generation rise. This slow ramp-up of energy gen-eration occurs for the full 10 billion years that the Sun generates all of its energy by fusion of hydrogen to helium.

The Sun's brightness increase is slow but continuous and inevita-ble. All stars like the Sun do it. The Sun has increased in brightness by about 30 percent in the last 4.5 billion years. The rise in bright-ness increases the intensity of sunlight illuminating planets. To put a 30 percent increase of solar brightness in perspective, if the Earth were moved to the orbit of Venus, the intensity of sunlight would increase by 50 percent. This would cause the oceans to be lost to space and create hellish conditions, similar to those that exist on Venus. The brightness growth is accelerating, and in 4 billion years it will be twice as bright as it was 4 billion years ago. Ultimately the nuclear "burning" process will move outward, as a shell of hydrogen surrounding a nearly pure helium core. At this point the Sun will enter what is known as the red giant stage, in which its surface will become cooler but its diameter will expand so much that its overall brightness will increase thousands of times. The slow brightening of the Sun during its middle age is less much less dramatic than the events that will occur when it becomes a red giant near its 10 billionth birthday. In its middle age it only brightens by a factor of two, but nonetheless this produces important stress on and change for the inner planets. On a "simple" Earth-like planet (if there were such a thing), the rise of solar heating by a factor of 2 would increase the surface by about 100°C. The Earth is not at all a simple planet, and it has many complex chemical and physical processes that influence its surface temperature besides just the brightness of the Sun.

The known rise in solar brightness has produced something of a mystery in Earth history called the "faint Sun paradox." In the past, the Sun was fainter and the Earth should have been cooler—so much cooler that that the oceans should have frozen. Aside from the short-lived "Snowball Earth" episodes 700 million years and 2.3 billion years ago, there is no evidence from the geological record that oceans were ever frozen for long periods of time. It appears that many factors have been involved in keeping the Earth's surface at a reasonably constant temperature in spite of significant evolution in solar brightness. In the Earth's past, the temperature rise has been accompanied by a host of different changes, including changes in land area, atmospheric composition, and volcanic activity. Some of these changes, such as the decrease in the abundance of carbon dioxide, a greenhouse gas, have compensated for increased solar brightness. As the Sun becomes brighter, there is less CO_2 and less greenhouse warming. Although the Earth has maintained habitability in the past, it will be not be able to do so indefinitely. For example, though this may seem counterintuitive, the CO_2 abundance is nearly zero, and it cannot reduce much further to compensate for increases in solar heating.

For all of its history, the Earth has been within the temperate zone of the solar system. This is the "right" range of distance from the Sun where an Earth-like planet can have surface temperatures that allow oceans and animals to exist without freezing or frying. This is called the habitable zone (HZ), and it extends from a well-known limit just inside the Earth's orbit, to a less-understood outer limit near Mars or possibly beyond. The HZ moves outward as the Sun becomes brighter, and in the future the zone will pass the Earth and leave it behind. The Earth will become Venus! The inner edge of the

[handwritten marginal note: Too simple~ clear that HZ is much larger than once presumed]

HZ is only about 15 million kilometers away, and it will effectively reach the Earth a half billion or a billion years from now (or less). After this time, the Sun will be too bright for animals to survive on Earth.

The final evolution of stars is reasonably well understood from a century of detailed research. The final stage of stellar lifetimes is short-lived, but this is only a relative term. For the Sun, its "normal state" (when it is similar to what we observe to today) lasts over 10 billion years, while the advanced state—the "red giant" phase—lasts less than a billion years.

The Planetary Thermostat

The steadily rising amount of energy hitting the Earth from the Sun would have long ago ended life on Earth—as it did on Venus (assuming that Venus ever had life)—except for one of the most important of all of the planetary life support systems: one that we can call the planetary thermostat. For more than 3 billion years (and perhaps 4 billion years), this system has kept the global average temperature of the Earth between the boiling and freezing points, thus allowing the most important requirement for life—liquid water—to continually exist on the surface of the planet for that immense amount of time.

The planetary thermostat is composed of three important subsystems: plate tectonics, the carbon cycle, and the carbonate silicate cycle. That the Earth has maintained a rather constant temperature through time is one of the major lines of evidence cited by Lovelock that the Gaia hypothesis is correct.

Imagine that it is not blood that circulates in your veins, but stone. A fantastic thought. But this was the analogy of one of the pioneers of geology, James Hutton. In the late eighteenth century Hutton used a scientific analogy based on the human circulatory system. According to historians of science, Hutton applied the metaphor of circulation to the cycle of rocks that he observed on Earth's surface. Today we know of a far more profound circulatory system—plate tectonics, a process that produces (among many other things) the movements of landmasses that we call continental drift. Of all integrated system processes, none is so important in maintaining the

Earth as we know it as plate tectonics, or continental drift. While at first glance it would seem that plate tectonics is a property solely of the solid Earth system, much evidence now shows that the atmospheric and hydrosphere systems are necessary to keep plate tectonics running in its current state.

The plate tectonics system is essential in maintaining surface temperatures at levels allowing the existence of liquid water on Earth. We might analogize plate tectonics with the physiological system that allows mammals and birds to maintain a constant body temperature, neither too hot nor too cold. Yet this is but one of the contributions of plate tectonics to the world as we know it. We can analogize the slowly moving continents and the enormous, molten convection cells that drive them as an enormous circulation system. But this system not only carries material from place to place, it changes material as well. The upward and downward movement of Earth material buries some material and liberates other material. It causes chemical changes as well, through new mineral formation, heating, and the liberation of gases. All of these aspects play a role in maintaining a constant temperature of the planet, and the mechanism for doing so will be described in more detail shortly.

The Carbon Cycle—the Transition from Organic to Inorganic and Back Again

If plate tectonics provides the largest arterial system for the planet, we can analogize several elements involved in that circulation as the blood. While many of these circulating elements or compounds, such as phosphates and nitrates, are essential for life, perhaps the most important element involved in cycling is carbon. The carbon cycle (fig. 3.1) is the main process for the regulation of long-term temperatures as well as atmospheric composition, and it is especially important in controlling future climate as the Sun bombards the Earth with ever more energy.

The rapid exchange of carbon, the crucial atom of terrestrial life, between inorganic compounds and organic compounds is crucial to life. Carbon not only is required for life to exist (and thus must be acquired during the lifetime of an organism to allow new cell growth

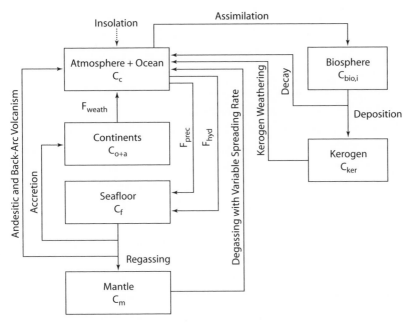

Figure 3.1. The carbon cycle. Arrows indicate directions of flow of carbon atoms. Source: Franck et al. (2006).

and repair) but is also, coincidentally, of overriding importance to the temperature of the Earth when it occurs as CO_2 in the atmosphere. Carbon dioxide is a "greenhouse gas," a gas that warms a planet's surface by absorbing radiant heat, also known as infrared radiation, and sending some of it back toward the Earth's surface, rather than allowing it to escape out into space. Methane (CH_4) and even water vapor are also highly effective greenhouse gases.

The movement of carbon from the inorganic to the organic world and back again has been described in many places, but my favorite passage comes from Lee Kump, James Kasting, and Robert Crane in their 1999 textbook, *The Earth System*. They begin with one CO_2 molecule, floating in the atmosphere. Over a decade or so it is wafted over and around the surface of the Earth by the moving, turbulent atmosphere, and over this time span it visits both the Northern and the Southern Hemisphere. During this time it might encounter a plant, and it passes into the plant itself through one of the small openings found in the leaves. Once within, it collides with other molecules and has its two oxygen atoms stripped away, to be replaced by hydrogen, nitrogen, and other carbon atoms through chemical bonding. In this way it is incorporated into the galaxy of atoms making up the material framework of the plant. Through this transformation we can say the particular carbon atom has been transformed from inorganic to organic.

The carbon atom exists within the organic framework through the summer, but as autumn comes, the leaf of which this carbon is a part falls from the tree and is soon buried in a thick blanket of other leaves and rotting vegetation. The leaf disintegrates in the fall rains and becomes incorporated into the soil among the myriad other organic molecules that had so recently been part of a green living leaf. Bacteria consume the organic molecules, and our particular carbon atom is transformed back into carbon dioxide by a chemical reaction initiated by the bacterium. Alternatively it may be consumed by an animal, and once again transformed into a CO_2 molecule that escapes into the atmosphere.

This history (with numerous variants) might be repeated up to five hundred times (the estimated average number of such cycles for a typical carbon atom) before a different fate occurs. In this case the soil containing our partially carbon atom is eroded and transported by moving water and eventually carried into the sea. There, it might again be consumed by an organism, but in this case let us assume that it escapes that fate and is buried by sedimentation. As more and more sediment falls onto this particular area of sea bottom, the carbon atom, still locked into a larger organic compound, is buried so deeply that it is now in a world of virtually no oxygen and thus exists in an environment that can no longer be perturbed by sediment-consuming animals such as polychaet worms and burrowing echinoderms. There are still bacteria down here, of course, and another likely fate might be transformation by bacteria back into CO_2, which might become dissolved in seawater but eventually be liberated as gas once again into the atmosphere. Yet a third fate is that this atom becomes entombed within sedimentary rock. The marine sediment in which it exists is now part of a sedimentary rock passage that underlies the ocean bed. Millions, tens of millions, hundreds of millions of years might pass, and eventually the sedimentary rock is thrust up into high mountains where it is eroded. The eroded sedimentary rock liberates its carbon, which bonds with atmospheric oxygen to form (again) inorganic CO_2.

The various fates of the carbon atom profiled above suggest that there are numerous "holding tanks," or reservoirs, where carbon is stored while it awaits its next incarnation. Some of these, such as the amount of carbon locked up as atmospheric methane, are small. Others, such as the amount locked up in sedimentary rocks, are many orders of magnitude larger. For long-term temperature, however, it is the relatively small volume of carbon locked up in the atmosphere that is of the greatest importance to biosphere health. Because of the disproportionate effect that tiny differences in carbon gas volumes exert on global temperature, even small perturbations in the inflow and outflow of carbon with the atmosphere produce relatively large swings in average global temperature.

It has been the relatively long-term steady state of atmospheric CO_2 levels that has been a key to the long-term habitability of the Earth. While there are seasonal imbalances between inflow (the processes of respiration and decomposition that cause carbon to be released from organic sources and enter the atmosphere as inorganic carbon) and outflow (the effects of photosynthesis, where atmospheric CO_2 is taken from the atmosphere by planet and converted into organic carbon), over the course of a year there is a steady state. But short-term variations in temperature also occur, causing a most familiar state of the Earth—its climate.

Long-term climate—and the maintenance of a global thermostat setting of between 0 and about 40°C over billions of years—is largely controlled by what has come to be called the silicate–carbonate geochemical cycle, and it is an integral part of the carbon cycle. This cycle involves the movement (transfer) of carbon to and from the crust and mantle of the planet, and it is accomplished by the plate tectonic system described earlier. This cycle, which involves living organisms, balances inorganic reactions taking place deep in the

Earth with interactions between the atmosphere and the surface of the Earth. It is this balance that keeps atmospheric CO_2 levels essentially constant—and hence keeps the Earth's surface temperature relatively constant—for the long timescale of geological time.

Two quite different processes are keys. The first is carbonate precipitation. If calcium is combined with carbonic acid (HCO_3) under correct temperature and pressure conditions, it can form calcium carbonate, the rock type also known as limestone. Limestone is one of the most common of all sedimentary rocks and is used widely by many types of organisms to help build skeletons of shell or even bone. The rate at which limestone forms on the surface of the Earth has important consequences for long-term climate.

The second major reaction involves the weathering of a class of rocks known as silicates. Weathering is the chemical or physical breakdown of rocks and minerals. When silicate rocks weather, the byproducts can combine with other compounds to produce calcium, silicon, water, and carbonic acid. In chemical terms the carbonate precipitation equation can be shown as:

Mix Ca^2 (calcium) with $2HCO_3$ (bicarbonate ion) and a chemical reaction will make $CaCO_3$ (limestone) and H_2CO_3 (carbonic acid)

The silicate weathering equation can be shown as:

Combine $CaSiO_3$ (rock silicates such as granites) with $2H_2CO_3$ (carbonic acid) and the chemical reaction will make Ca^{+2} (calcium ions) plus $2HCO_3^-$ (bicarbonate ion) plus SiO_2 (silica dioxide) plus H_2O (water)

These two chemical reactions combine in the following reaction:

$CaSiO_3$ plus CO_2 reacts to form $CaCO_3$ and SiO_2

The net result of this is that for each mole (a chemical term indicating a certain number of molecules) reacting, there is a net consumption of one mole of CO_2, which is buried on the ocean floor as limestone. The weathering of the silicate rocks thus eventually removes CO_2 from the atmosphere. This is of enormous importance to life. It is slight perturbations to the rates of those equations that will spell ultimate doom for plant life, and eventually for all life on Earth. If the Gaia hypothesis is correct, the role of organisms in this cycle should keep temperatures stable and optimal for life. Yet as we will see in the next chapter, such has not been the case. Life has repeatedly interfered with the planetary thermostat in important and nearly disastrous ways.

The most important element in *reducing* atmospheric carbon dioxide (which leads to global cooling) is the weathering of minerals known as silicates, such as feldspar and mica (granite has many such minerals within it). The presence or absence of plate tectonics on a given planet greatly affects the rates and efficiency of this "global thermostat." To reiterate the process, the basic chemical reaction is

$$CaSiO_3 + CO_2 = CaCO_3 + SiO_2.$$

By combining the first two chemicals in this equation, limestone is produced and carbon dioxide is removed from the system. The feedback mechanism at work here, first pointed out in a landmark 1981 paper by Walker, Hays, and Kasting, relates to the rates of weathering. Although weathering involves the reduction in size of rocks (big boulders weather into sand and clay over time), there is also a very important chemical aspect involved. Weathering can cause the actual mineral constituents of the rocks being weathered to change. Weathering of rocks containing silicate minerals (such as granite) plays a crucial part in regulating the planet thermostat. Walker and his colleagues pointed out that as a planet warms, the rate of chemical weathering on its surface increases. As the rate of weathering increases, more silicate material is made available for reaction with the atmosphere, and more carbon dioxide is removed, thus causing cooling. Yet as the planet cools, the rate of weathering decreases, and the CO_2 content of the atmosphere begins to rise, causing warming to occur. In this fashion the Earth's temperature oscillates between warmer and cooler due to the carbonate–silicate weathering and precipitation cycles. Without plate tectonics, this system does not work as efficiently. It also works less efficiently on planets without land surfaces—and much less efficiently on planets without vascular plants such as the higher plants common on Earth today.

Calcium is an important ingredient in this process, and it is found in two main sources on a planet's surface: igneous rocks and, most importantly, the sedimentary rocks called limestone. Calcium reacts with carbon dioxide to form limestone. Calcium thus draws CO_2 out of the atmosphere. When CO_2 begins to increase in the atmosphere, more limestone formation will occur. This can only happen, however, if there is a steady source of new calcium available. The calcium content is steadily made available by plate tectonics, for the formation of new mountains brings new sources of calcium back into the system in its magmas and by exhuming ancient limestone, eroding it, and thus releasing its calcium to react with more CO_2. At convergent plate margins, where the huge slabs of the Earth's surface dive back down into the planet, some of the sediments resting on the descending part are carried down into the Earth. High temperature and pressure convert some of these rocks into metamorphic rocks. One of the reactions is the carbonate metamorphic reaction, where limestone combining with silica converts to a calcium silicate—and carbon dioxide. The CO_2 can then be liberated back into the atmosphere in volcanic eruptions.

The planetary thermostat requires a balance between the amount of CO_2 being pumped into the atmosphere through volcanic action and the amount being taken out through the formation of limestone. The entire system is driven by heat emanating from the Earth's interior, which causes plate tectonics. But as we have seen there is more to this cycle than simply heating from the interior. Weathering on the surface of the Earth is crucial as well, and the rate of weathering is highly sensitive to temperature, for reaction rates involved in weathering tend to increase as temperature increases. This will cause silicate rocks to break down faster and thus create more calcium, the building block of limestone. With more calcium available, more limestone can form. But the rate of limestone formation affects the CO_2 content of the atmosphere, and when more limestone forms there is less and less CO_2 in the atmosphere, causing the climate to cool.

Here is a key aspect of the overall Earth system that helps refute either Gaia or Medea. If the Medea hypothesis is correct, we should be able to observe or measure a reduction of habitability potential (as measured by the carrying capacity, or total amount of life that can live on our planet at any give time) through time, or as measured by an observable shortening of the Earth's ability to be habitable for life in the future. For our own Earth, habitability will ultimately end for two reasons. The first of these is not Medean; it is a one-way effect. The ever-increasing energy output of our Sun, a phenomenon of all stars on what is called the main sequence, will ultimately cause the loss of the Earth's oceans (sometime in the next 2 to 3 billion years, according to new calculations). When the oceans are lost to space, planetary temperatures will rise to uninhabitable levels. But long before that, life will have died out on the Earth's surface through a mechanism that *is* Medean: because of life, the Earth will lose one resource without which the main trophic level of life itself—photosynthetic organisms, from microbes to higher plants—can no longer survive. This dwindling resource, ironically, (in this time when human society worries about too much of it), is atmospheric carbon dioxide. The Medean reduction of carbon dioxide will then cause a further reduction of planetary habitability because the CO_2 drop will trigger a drop in atmospheric oxygen to a level too low to support animal life. This is an example of a "Medean" property: it is because of life that the amount of CO_2 in the Earth's atmosphere has been steadily dropping over the last 200 million years. It is life that makes most calcium carbonate deposits, such as coral skeletons, and thus life that ultimately caused the drop in CO_2, since it takes CO_2 out of the atmosphere to build this kind of skeleton. Life will continue to do this until a lethal lower limit is attained. This finding is important: in chapter 8 I will show a graph that supports this statement. As pointed out by David Schwartzman, while limestone can be formed with or without life, life is far more efficient at producing calcium carbonate structures—a process that draws CO_2 out of the atmosphere—than nonlife.

53

HUMAN SURVIVAL

There is only one way out of the lethal box imposed by Darwinian life: the rise of intelligence capable of devising planetary-scale engineering. Technical, or tool-producing, intelligence is the unique solution to the planetary dilemma caused by Medean properties of life. New astrobiological work indicates that Venus, Mars, Europa, and Titan are potentially habitable worlds at the present time, at least for microbes, just as the Earth was early in its history. Did they undergo a reduction in habitability because of prior Medean forces? And certainly the cosmos is filled with Earth-like planets, based on both new modeling of still-forming solar systems and observations by the Butler and Marcy planet-finding missions. While the "planet finders" cannot yet directly observe any planet that is Earth-sized (a planet of this size is still too small for us to see with our current technologies), the orbits exhibited by some of the Jupiter- and Saturn-sized planets that can be observed suggest that smaller, Earth-like planets might exist there. Would Medean forces occur in alien life, as well as Earth life? If such life were Darwinian, the answer would be "certainly."

4

MEDEAN FEEDBACKS AND GLOBAL PROCESSES

The distribution of the elements on the planet was
initially controlled by physical and chemical processes,
but biological processes have been at work in affecting
the chemical dispersal ever since life first appeared
about 3.5 billion years ago.

—M. Jacobson et al., *Earth System Science*, 2000

One of the fundamental findings of Earth system science has been the discovery of numerous "feedback" systems—where a given environmental change cycles through various systems and ultimately produces further change. James Lovelock noted these early in the history of the Gaia hypothesis; one of the predictions of the various Gaia hypotheses is that biological feedbacks—in which life plays an important part in the overall system and its effects—should be dominantly "negative." For instance, a negative biological feedback for planetary temperature would mean that rising temperatures would eventually cause the feedback system to bring about a subsequent lowering of temperatures, or that an initial decrease of atmospheric oxygen should, through the feedback system, cause an eventual, subsequent rise in oxygen. In this way, conditions are kept fairly stable. In the previous chapter, in which I defined the Medea hypothesis, I proposed the opposite—that biological feedback systems affecting the survivability of life are overwhelmingly positive, and even when negative (as in some of the minor, biological responses to CO_2, described in more detail below), they are almost inconsequential and are overwhelmed by other feedbacks that are positive.

There is another aspect of environment-enhancing feedback. If it occurs, it is intrinsically destabilizing, as noted by James Kirchner (p. 404), who stated relatively early in the formation of Gaia hypothesis:

> Organisms that make their environment more suitable for themselves will grow, and thus affect their environment still more, and thus grow still further. This is positive feedback, not negative feedback. Negative feedback arises when a growing population makes its environment less suitable for itself, and thus limits its growth. Environment-enhancing feedbacks are destabilizing; environment-degrading feedbacks are stabilizing. The Gaian notion of environment-enhancing negative feedbacks is, from the standpoint of control theory, a contradiction in terms.

This is what Ward is arguing!

In this chapter I discuss some examples of positive feedbacks that I would classify as "Medean," as well as introducing some processes resulting from life itself that are anything but beneficial to other life.

FEEDBACK SYSTEMS

Climate and climate change are, and surely always have been, primary determinants of the distribution and probably abundance of life on Earth. Variations in temperature and water availability, both largely determined by climate, are certainly primary determinants of the ranges of various organisms: there are no palm trees in the Arctic or water-loving plants in the desert, among innumerable examples. Thus it should be expected that, if organisms are somehow able to increase habitability by optimizing (as in Optimizing Gaia) or even regulating conditions for themselves (Regulating Gaia), they would do so through some aspect of climate change or change in atmospheric gas inventory, and that the mechanisms involved would be negative feedback systems. So what would be the inverse of this—the Medean prediction? Surely it would be that feedbacks are either positive (worsening conditions) or at best neutral (no change). In a 2003 review article on climate, James Kirchner, whose writings and thoughts have had a major influence on the conclusions of this book, explicitly noted the realization that there are indeed situations where life does not better the environment for itself but in fact makes

Who is he?

things worse: "Destabilizing feedback is often presented [by supporters of the Gaia hypothesis] as an aberration that arises during the breakdown of regulatory mechanisms (e.g., Lovelock and Kump, 1994), rather than an intrinsic characteristic of many biologically mediated processes."

Just how pervasive are these "destabilizing"—or positive—feedbacks? Let us look at some examples illustrating that the latter seems to fit known history and current observations. Nowhere is this better illustrated than in the effect that rising CO_2 is having on world climate.

TEMPERATURE (FROM CO_2)

The current rise in atmospheric CO_2 (and CH_4) is and will continue to be one of humankind's great challenges. That this is taking place at all is also a challenge to Homeostatic Gaia, but the systems are multiple and anything but simple. The feedback systems involving just CO_2, for instance, include the following, as listed in the review by Kirchner (2002, pp. 395–96):

1. Increased atmospheric CO_2 concentrations stimulate increased photosynthesis, leading to carbon sequestration in biomass (negative feedback).

2. Warmer temperatures increase soil respiration rates, releasing organic carbon stored in soils (positive feedback).

3. Warmer temperatures increase fire frequency, leading to net replacement of older, larger trees with younger, smaller ones, resulting in net release of carbon from forest biomass (positive feedback).

4. Warming may lead to drying, and thus sparser vegetation and increased desertification, in mid-latitudes, increasing planetary albedo and atmospheric dust concentrations (negative feedback).

5. Higher atmospheric CO_2 concentrations may increase drought tolerance in plants, potentially leading to expansion of shrublands into deserts, thus reducing planetary albedo and atmospheric dust concentrations (positive feedback).

6. Warming leads to replacement of tundra by boreal forest, decreasing planetary albedo (positive feedback).
7. Warming of soils accelerates methane production more than methane consumption, leading to net methane release (positive feedback).
8. Warming of soils accelerates nitrous oxide (N_2O) production rates (positive feedback).
9. Warmer temperatures lead to release of CO_2 and CH_4 from high-latitude peatlands (positive, potentially large, feedback).

If science were somehow democratic, this would be a clear vote by nature that either Homeostatic Gaia is not at work with regard to rising CO_2 or we have described the systems wrongly, since there are seven positives to two negatives. A more accurate way to deal with this is to ask what percentage of the CO_2 entering the atmosphere is taken back out of the atmosphere through some sort of sequestering—for example, by the oceans. Here too, it appears that a small fraction of the atmospheric CO_2 is being taken back out, and organisms are involved in only a small amount of this (for the greatest sequestration factor is the dissolution of CO_2 in seawater, which is abiotic rather than biotic in action). In fact, according to the pioneering work of Lashof et al. (1997), it seems as if organisms are actually *amplifying* the effects of global warming through an increase of atmospheric CO_2.

Is there data that can help us choose between these alternatives—whether organisms will significantly reduce global temperatures by reducing CO_2 or will instead either produce a negligible reduction or actually cause global temperatures to increase? In fact, there is an unparalleled data set for the recent and deep past that is highly relevant: ice cores taken from thick ice sheets, mainly in Greenland and in the Antarctic region near Lake Vostok. By analyzing the gases that were trapped in the ice over the course of many years, the atmospheric composition and global temperature of the ancient past can be determined.

The Greenland and Vostok ice core records, as well as historical observation (for example, from the Hawaiian CO_2 observatory), clearly show that atmospheric CO_2 has been rising well above Holocene and even Pleistocene values during the past two centuries. But why is this happening? One of the original tenets of James Lovelock was that the atmosphere of the Earth was strongly controlled (regulated) by organisms. Thus, one would predict that biological processes should tightly regulate the composition of the atmosphere and therefore keep it relatively stable; in other words, we should not see appreciable changes in oxygen or carbon dioxide levels over time. This is clearly not true. Both long-term and shorter-term changes in the levels of these biologically important gases show that the atmosphere not only can change but can, under certain conditions (such as when there are ice caps on Earth), change quickly.

In the short term, the ice core records indicate that there has been a 35 percent rise in atmospheric CO_2 since preindustrial times, *but rates of carbon uptake into the biosphere have accelerated by only about 2 percent.* Additionally, in response to the Gaia theory prediction that atmospheric CO_2 should be more sensitively regulated by terrestrial ecosystem uptake (which is biologically mediated) than by ocean uptake (which is primarily abiotic), both processes are about equally insensitive to atmospheric CO_2 levels.

Viewed in these quantitative terms, the coupling between atmospheric CO_2 and carbon uptake by the biosphere is weak, consistent with Lashof's (1989) estimate of a negative feedback gain of only −0.02. In other words, the feedback system that should reduce carbon dioxide is close to neutral, rather than having a significant effect on CO_2 reduction. The fact that the oceans take in as much CO_2 through abiotic means as does the plant-rich terrestrial biosphere is certainly not in line with the hypothesized biotic feedback system, which is supposed to limit CO_2.

In fact, the Vostok ice core record shows that, to the extent that the Earth system regulates CO_2, CH_4, and dimethyl sulfate (DMS) in the atmosphere, all three of these planetary "thermostats" are hooked up backwards, functioning to make the Earth cooler during glacial periods and warmer during interglacials (Petit et al. 1999), and they apparently *destabilize* Earth's climate on timescales of hundreds of thousands of years. This demonstrates that atmospheric CO_2

and CH_4 are *not* tightly regulated by the Earth system, even though both are important controllers of the Earth's climate, and even though CO_2 participates directly in the most fundamental processes of life. *Thus the failure of the Earth system to tightly regulate atmospheric CO_2, at least on human timescales, is another empirical refutation of the Gaia hypothesis.*

The last of these supposed thermostats, dimethyl sulfate, was originally thought to act as a global thermostat (Charlson et al. 1987); only later did it become apparent that it serves to make the Earth cooler when it is cool and warmer when it is warm. This conclusion came from multiple authors (Legrand et al. 1988, 1991; Kirchner 1990; Watson and Liss 1998). It was thought that phytoplankton produce DMS, which itself is a nucleating agent for clouds. The system was thought to be a negative feedback on temperature, a kind of marine biological thermostat. An increase in plankton formation would lead to more DMS, but DMS would cause greater cloud cover, and through this increased albedo the temperature of the planet would drop. But later work showed that DMS is really produced mainly by dust, not phytoplankton, and that it seems to act exactly opposite to the way it was originally posited: this so-called thermostat actually increases temperatures when temperatures increase—a positive feedback. This is a Medean result.

Gaia theory also predicts that biological feedbacks should make the Earth system less sensitive to perturbation, but the best available data suggest that the net effect of biologically mediated feedbacks will be to amplify, not reduce, the Earth system's sensitivity to anthropogenic climate change. Measuring the degree of perturbation is daunting and fraught with potential error, so I do not deal with it here. I predict, however, that it will ultimately be shown that biological effects increase rather than dampen climatic, oceanic, and atmospheric chemistry changes that are deleterious to life.

Given the apparent pervasiveness of these destabilizing biological feedbacks (positive feedbacks), it is appropriate to question whether the Earth system has been stabilized *by* biological feedback processes or *in spite of* them—the latter a Medean result. In fact, the scientific and policy-making communities have not adequately considered the risk that biogeochemical feedbacks could substantially amplify global warming.

]Wish I
]were
one!

BIOLOGICALLY PRODUCED LONG-TERM PERTURBATIONS IN OXYGEN AND CO_2

At this point we need to start delving deeply into the Earth's geological past. To do this, we must first talk about time.

"Precambrian" time is broken into three major divisions, named (from oldest to youngest) the Hadean, Archean, and Proterozoic eons. The Hadean was the time before life and any sort of abundant rock record. The Archean began with the first appearance of life and a rock record but ended not with any biological event; instead, it concluded with a series of physical changes to the Earth. The succeeding Proterozoic was a time dominated by microbes, but near its end the first animals evolved. The boundary between the Proterozoic and the Cambrian Explosion marks the succeeding Paleozoic, when skeletonized animals appeared in large numbers for the first time. The Hadean, Archean, and Proterozoic are thus long intervals of time with few definable intervals.

New information about both oxygen and carbon dioxide through time has recently become available, and for details and processes of these changes the reader should consult Robert Berner's excellent, if technical, book *The Phanerozoic Carbon Cycle* (2004) or my own *Out of Thin Air* (2006). The amount of carbon is a function of carbon dioxide levels in the atmosphere, and this number can be estimated through time. The CO_2 in the atmosphere was originally from volcanic and deep Earth sources.

Robert Berner of Yale University has been the leader of studying CO_2 content over time. His (and various colleagues') goal was to calculate the amount of carbon dioxide in the atmosphere in the past using a mathematical model, for there is no way to measure this value directly from ancient rocks. In essence, they were looking at the balance between continental weathering of carbonate and calcium magnesium silicate rocks (which, as mentioned above, liberate Ca^{+2} and eventually remove CO_2 from the atmosphere through the formation of limestone in the sea) and the input of new CO_2 back into the atmosphere from volcanic emission, itself related to the subduction of seafloor limestone brought about by plate tectonics. In this model, they had to worry about four main variables. The first is continental land area. Because the rate of weathering is related to the amount of land that can be weathered, the size of the continent through time will have a decisive effect on CO_2 levels. With more land area on the planet, there are more silicate rocks to weather—and hence a great ability for the feedback system to remove CO_2 from the atmosphere.

The second variable is the rate of seafloor spreading. Spreading rates seem to be themselves related to the amount of heat emanating from deep within the Earth. With more such heat, there is more volcanic activity—and hence more volcanically generated CO_2 being pumped back into the atmosphere.

The third variable is the weathering rate. As temperature goes up, so too does the weathering rate, which affects the feedback system. Finally, the fourth factor is a chemical one: the concentrations of calcium, bicarbonate, and calcium carbonate in the Earth, atmosphere, and ocean systems are a function of the amount of CO_2 present in the atmosphere.

Having so many factors to model requires a complex mathematical solution. The most "primitive" of the first mathematical solutions (which was dubbed the BLAG model after its authors, Berner, Lasaga, and Garrels) required the simultaneous solution of eight differential equations. In its first incarnation, a rough prediction (for that is all models can do—they do not measure past CO_2 values, but only give us some idea of what they may have been), a curve of CO_2 for the past 100 million years was generated.

Over a ten-year span into the 1990s, the BLAG model evolved. Its authors, realizing that the many simplifications of the first model could yield only very crude predictions, tried to improve the modeling mathematics by increasing the sophistication of the assumptions, and by including new mathematical expressions. For instance, in the new and improved model (named GEOCARB), new wrinkles included the incorporation of the increasing luminosity and energy flux of the Sun through time, adding better information about the rate of seafloor spreading for the last 150 million years (and thereby better understanding the rate at which CO_2 was being transmitted into the atmosphere through volcanoes), and, perhaps most important, incorporating new expressions modeling the effects of biology on weathering. Since the rate of rock weathering is the key component to the system, the dawning understanding through the 1990s of the importance of biology on weathering required that this be taken into account. The most important breakthrough was the recognition that the invasion of the land by plants over the past 500 million year must have drastically changed weathering rates—and hence the cycle of CO_2 among land, air, and sea. The primacy of biology itself invaded the mathematicians' turf.

All of the models described below are possible due to the understanding of how aspects of the Earth systems interact with themselves and with the external environment. There are both positive and negative influences. In a positive feedback, an increase of energy or rate within a given system causes another to increase as well. In a negative interaction, an increase in rate causes a decrease within an interacting system. Lastly, there are neutral interactions, where an increase or decrease has no effect on another system. These interactions can be illustrated with a flowchart (fig. 4.1).

As shown in the figure, a wide variety of interactions take place—thirteen, to be precise. For instance, as solar luminosity increases, mean global temperature increases as well, which both increases the rate at which the silicate rock weathering takes place and reduces biological productivity. Increasing silicate rock weathering decreases the amount of atmospheric carbon dioxide, whereas increasing biological productivity increases the rate of silicate rock weathering (as does increasing the amount of atmospheric carbon dioxide). Biological productivity is increased by the amount of atmospheric carbon dioxide. Increasing geothermal heat flow increases spreading rate, which causes faster silicate rock weathering. Increasing the rate of continental growth increases rock area and also causes the rate of silicate rock weathering to increase.

After plugging the various values into this model and feeding all of them into various computers, Berner and his colleagues ended up

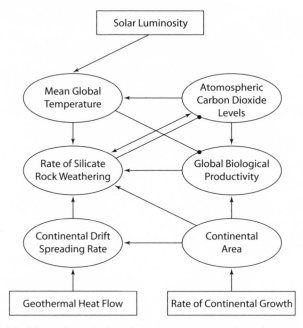

Figure 4.1. Models used to calculate future temperature and productivity.

TABLE 4.1

Processes affecting the weathering rate of silicate and carbonate rocks as well as organic matter on continents

    Topographic relief as affected by mountain uplift

    Global land area

    Global river runoff and land temperature as affected by continental drift

    Rise of vascular land plants

    Rise of angiosperm (flowering) plants

    Changes of global temperature brought about by evolution of the Sun; changes
        in atmospheric CO_2; rate of mineral dissolution

    Enhancement of plant root activity due to fertilization by CO_2

Processes affecting the rate of thermal degassing of CO_2 from the subsurface due to volcanism and metamorphism

    Changes in seafloor spreading rate

    Transfer of $CaCO_3$ between shallow- and deep-water areas in the sea

---

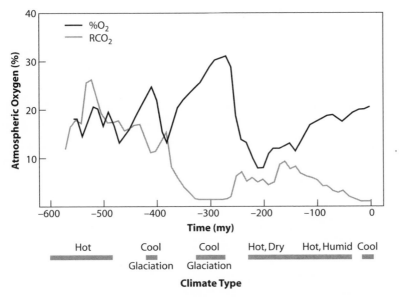

Error bars ??

Figure 4.2. Carbon dioxide (and oxygen) curves through time, with climate regimes indicated below the curves. High levels of CO_2 correspond to greenhouse warming. RCO_2 is the ratio of the mass of CO_2 to that of the present.

✓ Steve J

with the graph shown in figure 4.2. The Berner graph, as calculated for the last 600 million years (roughly the time that animals and higher plants have existed on planet Earth) shows several interesting trends, the most important being an overall, long-term decrease in CO_2. At the start of the studied interval, a time interval known as the Cambrian period, CO_2 levels were about fifteen times higher than present-day levels, and then, over the subsequent 100 to 150 million years, they gradually increased through a series of fluctuations to more than twenty times present-day values. But about 400 million years ago the most remarkable thing happened: CO_2 levels dropped markedly. The reason for this drop seems clear. The time interval of about 400 million years ago coincides with the rise of vascular land plants.

As land plants began to cover the planet with the first sparse twiggy forms, soon evolving into higher shrubs and eventually reaching into the sky as the tallest of trees, enormous changes affected the planet. Great quantities of carbon began to be locked up into the land within rotting vegetation, and eventually coal and oil. Soils became deeper and richer. And as the green spread, the fine balance between the amount of carbon held in the atmosphere and that held in the soils, oceans, and rocks of the planet began to change. Carbon dioxide levels began to drop as more and more plants sucked it from the atmosphere, and as plants began to increase the weathering rates of silicate rocks, thus allowing ever more limestone to form. The history of CO_2 abundance over the last 100 million years has been one of overall decline. Much of this decline may have been driven by tectonic events, most notably the geological uplift and subsequent weathering of the Himalayan Mountains. Because this largest of the Earth's mountain ranges is composed largely of silicate rocks, and because its extraordinary uplift (due to the chance collision of the Indian tectonic plate with Asia) created the thickest continent crust on the planet, this single event seems to have markedly changed atmospheric CO_2 composition, and with it, the Earth's climate. As the Himalayas weathered, the liberation of calcium and silicate ions caused the various carbonate cycles to remove atmospheric CO_2 faster than the volcanic system could replace it. Over the last 60 million years, this event, coupled with the further spread of plant life across the land, drove the levels of CO_2 to a historical low. Over the past 500 million years that amount has fluctuated, but in its troughs it has never reached a point where the existence of plants was threatened. However, there has been a long-term drop.

Another aspect of CO_2 and oxygen relates to mass extinctions. As we will see, specific mass extinctions appear to have been coincident with times of either low oxygen or increasing CO_2. I have recently used new data from Berner's work to demonstrate this remarkable relationship, shown in figure 4.3, where the bars are the times of mass extinctions.

While increases in CO_2 are largely abiotic (from flood basalt volcanism), the changes in oxygen are largely linked to biology. For example, the highest rise in oxygen, which occurred during the Carboniferous period of 350 to about 300 million years ago, took place as a result of the rapid burial of organic carbon. In this case, the

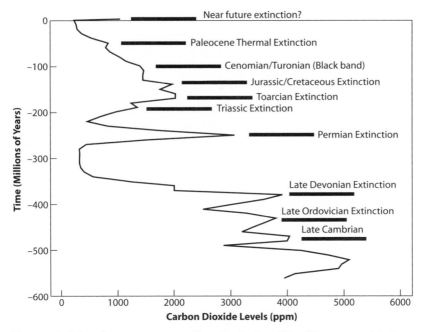

Figure 4.3. CO_2 values in parts per million for the last 600 million years, with times of mass extinction shown.

carbon was stored in the newly evolved trees, whose lignin proved resistant to breakdown and decomposition by the microbes of the time. The rapidly buried trees removed a large amount of material that would otherwise have bound with oxygen (thereby quickly reducing CO_2 levels in the atmosphere). The opposite effect—drops in oxygen—came from the other end of the spectrum: too much so-called reducing material available to bind with oxygen. We will see in the next chapter that both caused major reductions in biomass—that is, Medean results.

BIOTICALLY MEDIATED PLUNDERING

Another aspect of life that is Medean in nature is a process referred to as biotically mediated plundering. This occurs when one type of organism uses resources in such a way that there is inhibition of other organisms, or, in the case of oceanic eutrophication (described in more detail below), such that the overuse can actually lead to extinction of the users. A good example is the "biotic plunder" of nutrients from the surface ocean, as plankton take up nutrients, die, and sink (Volk 2002). This biologically mediated process has created a biological desert over most of the world's oceans, constituting over half of the Earth's surface area. Only the nutrient starvation of the plankton limits biotic plunder by plankton themselves.

Nes
feedback

Another example comes from organisms that hoard resources for themselves, thus depleting their environments. Resource hoarding can often increase fitness, and thus, through natural selection, resource-hoarding organisms can become more common (and thus resources become scarcer); the result is that organisms that efficiently hoard resources will gain an ever-growing advantage over those that do not.

OTHER MEDEAN EFFECTS

There are other biological effects that adversely affect the biosphere and its organisms. Two of these are oceanic eutrophication and direct poisoning.

Oceanic Eutrophication

This process, long known to occur in lakes in terrestrial settings, has only recently been hypothesized to occur in the oceans as well. In fact, this mechanism is now thought to have caused one of the five largest mass extinctions in Earth history, the event that occurred about 360 million years ago in the Devonian period.

As in lakes, oceanic eutrophication takes place when blooms of organisms (usually phytoplankton) undergo a boom and bust life cycle in response to sudden, anomalous increases in nutrients. It seems ironic that too much life is a dangerous situation for a community of organisms. As a result of the increase in nutrients, the population of organisms that feed off of these nutrients swells, sometimes to high levels, choking the surface regions in the case of plankton. However, if the nutrient supply is then reduced, the overpopulation of plankton will starve and largely die off. Many of these dead bodies fall downward and rot, and the rotting uses up oxygen in the water—oxygen that is necessary for life of the still living. It is thus the subsequent decomposition of this great mass of organic carbon that causes the problem: as a result of a chain reaction sparked by the bloom and subsequent death of a large population of organisms, all of the oxygen in the water column will be used up by the decomposition of the dead bodies, killing off various other organisms in the surface ocean.

Here is how this mechanism is thought to work. These events of the past have left a record in the rocks, one that can be deciphered by careful geological observation. Often there is a geological transition from carbonates (limestones) to black mud rocks—the latter usually indicative of sedimentation in oxygen-deficient settings, and also often accompanied by a mass extinction of organisms that lived there during limestone formation. Often there is a reduction of limestones, their place taken by the mud rocks (the fossils in these rocks change as well). Before the oxygen depletion event, most of the animals and plants inhabiting the sea bottom are photosynthesizing (modern corals have microscopic plants called dinoflagellates in their flesh that undertake photosynthesis, to the benefit of the coral). This changes to a benthic assemblage characterized by scavengers, deposit feeders, and carnivores. An upwelling (the rise of bottom waters to the surface, caused by a variety of submarine currents) of nutrient-laden deeper waters causes the nitrogen and phosphorous availability in the water column to skyrocket during these intervals, leading to an explosion in primary productivity. Because tropical to subtropical organisms in the Devonian period most likely adapted to low-nutrient, clear-water conditions, the proposed events would have had a major impact. The massive blooms of phytoplankton would have eclipsed surface water clarity, killing off plankton and communities of swimmers alike that had been adapted to clear water (which allows photosynthesis). In the lower water column, the subsequent widespread oxygen deficiency would have demolished benthic communities, and the bloom would indirectly have caused this. The subsequent death and fall onto the bottom of the many bodies of the phytoplankton would have used up oxygen in the bottom part of the seawater column, causing the ultimate demise of carbonate-producing benthic organisms.

These "eutrophication" events are just now being discovered from the ancient rock record. They have long been known from the study of modern lakes and fjords, however. In both cases, following a eutrophication event, there are fewer species, but there are also far fewer organisms and less biomass in general, since a replacement of animals and plants low in the food chain by organisms that are heterotrophic (they need living matter as food) results in a net drop in biomass. This is a Medean result.

The eutrophication event at the end of the Devonian, described here, will be profiled in the next chapter. As I will show, based on new results of my own research group, it may not have been the only case where this kind of event took place.

Poisoning

Direct poisoning of the environment is also the hallmark of many kinds of organisms. Some of these toxins are to inhibit other competitors. Australian eucalyptus trees are examples of this—their leaves and bark are highly toxic, and when spread around the tree trunk where they originated, this eucalyptus material effectively poisons any ground cover and then inhibits germination of invaders.

A second kind of poisoning is caused by the so-called red tide organisms, which produce and release toxins into the seawater that kill off all life around them.

A third kind of poisoning comes from the production by microbes of the deadly compound hydrogen sulfide (H_2S). This material is produced as a byproduct of several kinds of bacteria, forms that live only in low-oxygen seawater. As we shall see in the next chapter, this mechanism, causing a change in the "chemocline" (the boundary between deep, anoxic bottom water and shallower, oxygenated water), has only recently been discovered through the research of Lee Kump and his colleagues at Pennsylvania State University. There is a dose of irony here, as this is the same Lee Kump who authored with James Lovelock one of the strongest theoretical arguments for the existence of Gaia phenomena—the Daisyworld model. Yet here is Kump now illustrating a process that may have been one of the most efficient and encompassing killers ever produced or mediated by organisms. The H_2S release may have been responsible for the majority of mass extinctions, and it certainly could happen again. — ∴ Gaian ? |

There are many more examples of what might be called Medean phenomena. However, more important are the examples of these events in Earth history that I will profile in the next chapter.

SCALE ?

5

MEDEAN EVENTS IN THE HISTORY OF LIFE

> The disappearance of large numbers of species can ac-
> curately be called destructive and any selection that
> chiefly eliminates and seemingly does not allow the
> evolution of resistance to mass extinction is clearly
> coarse grained.
>
> —George McGhee, 1989

This chapter presents a list of events that, combined, provide abundant evidence that effectively refutes the Gaia hypothesis. This evidence does not "prove" the Medean hypothesis; proof is difficult to nigh impossible in science. But as it will be the last one standing, the evidence presented should certainly strengthen its acceptance. First I describe a series of episodes from Earth history, each producing events that should not have occurred if the Gaia hypothesis is correct. Then I look at the long-term history of life on this planet in terms of our best estimates of biomass through time. Using this as one way of deeming biotic success, the results will be examined in terms of the two hypotheses.

Each of the events below represents an episode in Earth history when some aspect of life threatened other life and/or itself, usually the former. The events are roughly arranged in chronological sequence.

DNA Takeover, 4.0(?)–3.7(?) Billion Years Ago—the First Medean Event?

Let us begin this chapter with a first proposal about what will be called Medean events—life-driven episodes that result in a drop in

[handwritten margin notes:]
A refutation of the Balance of Nature — long rejected in Ecology;
BUT ≠ stochastic; both occur; depends on context, scale
Analogy to body — focus on disturbance or recovery?
So life periodically causes death/extinction OK — BUT then life recovers/persists.

diversity of abundance of subsequent (later generation) life. These events—in reality, extinction events—are listed in temporal order, oldest to youngest. Unfortunately, unlike the later events, which are based on data from stratal and fossil evidence, this first Medean event is no more than an educated guess. Many of us believe it to be true, although since there is currently no way to scientifically test this hypothesis, it must remain an educated guess, and no more.

Not science, why include?

Over the past several years, biochemists conducting experiments to find possible alternatives to our familiar Earth life have attempted, and in some cases have succeeded, in producing DNA with exotic "languages," obtained by changing the number of nucleotides used to code for a specific amino acid. Experiments by various biochemists show that DNA could indeed come in many varieties of "languages"—for example, as detailed above, by using a different number of nucleotides to code for a particular amino acid. Thus, it is possible—indeed, I believe, likely—that early DNA life on Earth might have come in a variety of forms, perhaps all slightly different from our now familiar DNA. If so, it is probable that separate kinds of DNA competed against each other. There seem to be two possibilities—either that our current version proved competitively superior to the others, or it simply was the first to evolve from an RNA world. It is hard to believe that our complex variety of DNA appeared fully formed, without a competing cohort of slightly different versions. If such was the case, it is probable that there would have been intense competition between each of the kinds of DNA, competition that would be inherently Darwinian. The suppression of other kinds of life would have followed, and if so, this would have been a Medean event, in fact, *the first example of a Medean event: the takeover of the single kind of life.* Elsewhere I have followed other evolutionary biologists in supposing that the highest diversity of life, the most basic kinds of life, not species, was surely early in Earth history. Since that time it is probable that we have had but one kind of life.

Later in this book we will look at various kinds of mass extinctions. But as catastrophic as they were, these all occurred among DNA life, and mainly the higher, more complex kinds of that life, such as animals and higher plants. In reality, the greatest mass extinction of all may have been the first, the DNA (as we have it now, anyway) incurred mass extinction, and this event was caused by life—Darwinian, and hence Medean, Earth life.

The Methane Disaster, 3.7 Million Years Ago

The early Earth was a place vastly different from the planet we live on. What was the nature of the oceans and atmosphere when life first appeared, and did that life have any subsequent effect on the oceans and atmospheres?

To get a view of what the chemical nature of the early Earth may have been, we have only to turn to Titan, the largest moon in our solar system. Titan has a peculiar atmosphere, currently unique in our solar system, but one that any, or all, early Earth-like planets might necessarily have: it has a thick atmosphere of methane smog. The methane atmosphere hypothesized to have been present on the early Earth may have been a byproduct of the first life, a waste product of its earliest metabolism. Life was present as a series of oil-like slicks and stacked bacterial layers and sediment, called stromatolites. Although individually small in size, these microbes became globally abundant and in so doing began to change the planet—or, more accurately, to poison the planet. This is the conclusion of the eminent atmospheric scientist and astrobiologist James Kasting of Penn State. His new work describes how the formation of methane-producing life on Earth nearly ended the saga of life on our planet in its earliest epochs by creating a cold buffer on the surface of the planet. The formation of the methane haze took an already cold world (the Sun was more than 30 percent less energetic) and added a layer of clouds for the first time that reflected heat back into space. But for the very high volcanic heat flow on our planet, this type of condition made the planet much worse for life's survival. Had the Earth been even slightly farther out in space, the planet would have reached a temperature too cold for any kind of current Earth life. Whether a different kind of life, such as the possible ammonia life described in chapter 3, would have evolved is unknown. In any event, this byproduct of cooling clouds is not the predicted course of events under the Gaia hypothesis. This is the first test that shows how earliest life nearly ended its own history through its formation.

The First Rise of Atmospheric Oxygen, 2.5 Billion Years Ago—Chemical Weapon of Mass Destruction

The current idea about the origin of our oxygen atmosphere is that small amounts began to appear from then brand new, photosynthetic microbes some about 3 billion years ago. Until that time all life had been anaerobic, utilizing primitive types of photosynthesis that did not release oxygen. But the evolution of oxygen-releasing photosynthesis led to the accumulation of free oxygen. This near suicide is one of the most astounding examples of the Medea hypothesis. My understanding of this episode comes from the important new findings recently published in the *Proceedings of the National Academy of Science* by Robert Kopp and Joe Kirschvink of the California Institute of Technology. Their work on manganese fields in South Africa indicates that cyanobacteria, the microbes that caused the sudden appearance of oxygen, did not evolve until hundreds of millions of years after their supposed appearance. Oxygen caused a massive mass extinction on planet Earth: the biovolume of life on the planet plummeted. This is a Medean result. Only the children of the bugs that could tolerate oxygen—and the cyanobacteria that learned to make it, and the bugs that later learned to breathe it—would thereafter enjoy the sunlight.

The First Global Glaciation, 2.3 Billion Years Ago—Life Causes the First Snowball Earth

Joe Kirschvink made one of the great scientific discoveries of the late 1990s. He was the author of the now-accepted idea called Snowball Earth. This hypothesis proposes that the first global Ice Age, which took place between about 2.32 and 2.22 billion years ago, was so severe that the Earth's oceans froze over completely—with only heat from the planetary core allowing some liquid water to exist under ice more than a kilometer thick—and in so doing nearly extinguished life on Earth. This episode certainly reduced the amount of life on our planet by many orders of magnitude. And the event was Medean. The same photosynthetic microbes described above increased in number to the point that they removed most greenhouse gases, including carbon dioxide and methane. Since methane is a powerful greenhouse gas, and since the Sun (and its energy output hitting the Earth) was notably weaker at the time, temperatures plunged—this time not due to clouds, as in the first event when earliest life formed, but due to the loss of heat-cloaking greenhouse gases. Volcanoes, the main source of carbon dioxide, could not provide enough to maintain the level of greenhouse gases in the atmosphere that was needed to keep the Earth warm, and the planetary thermostat broke down. As a result, the polar ice caps expanded, covering a large part of the surface with glaciers. At that point, global cooling accelerated because ice, being white, reflects heat back into space much more efficiently than materials of darker colors. Noted Harvard geologist Paul Hoffman began working on the Snowball Earth phenomenon soon after Kirschvink published his first hypothesis paper, noting that if more snow reflects more sunlight back into space, this soon causes runaway glaciation. The ocean freezes over. There is mass mortality, and there are few survivors. But this third Medean episode was perhaps as close to planetary sterilization as we have ever gotten. It was brought about through the action of organisms and was thus a Medean action.

The Canfield Oceans, 2–1 Billion Years Ago (?)

One of the most perplexing of all questions is why it took so long for animals and higher plants to evolve from simpler ancestors on the ancient Earth. Surely the number of evolutionary steps (first to the eukaryotic grade, then to multicellular), while composed of many separate substeps, was not so difficult or time-consuming that it took literally billions of years. Yet we had life some 3.7 billion years ago, but we do not see true animals until less than 0.6 billion years ago. Even multicellular plants were not present in abundance until a billion years ago. Why so long? Recently it has been proposed that the evolution of complexity was stifled by life itself, in the form of what we call Canfield oceans. This was Medean.

Handwritten margin notes:

Medean = ?

Life interfering w/ itself

(long known & accepted at smaller scales)

What ended it? (Steve).

So anytime life (biomass or diversity) is not maximal due to biotic factor we conclude Medean — seems weaker or obvious

Geochemist Don Canfield, himself a student of the same Robert Berner of Yale who discovered the changes in Earth's carbon dioxide and oxygen records, has changed our view of the ancient oceans and atmospheres. Canfield and Berner published seminal papers about the atmosphere and oceans in the times of stratified, anoxic oceans, oceans characteristic of the long Precambrian (and also found during the periods of mass extinction to be profiled below). Their discovery, first using stable isotopes of sulfur and later confirmed by modeling, was that over long periods of time the anoxic oceans were of two kinds—one was simply without oxygen, but the other, while also anoxic, was filled with a veritable poison, the nasty gas known as hydrogen sulfide. So different was this kind of ocean that the geological fraternity named it the Canfield ocean after its prime discoverer.

Anoxic oceans following the evolution of life allowed massive quantities of reduced carbon to build up and eventually get buried. In some cases, however, *even in the absence of dissolved oxygen*, the reduced carbons do get oxidized, and the resulting product is hydrogen sulfide, or H_2S, gas. As any poor victim of freshman college chemistry knows, H_2S is the stuff that gives rotten eggs a bad name, and for good reason. This highly toxic stuff is a lethal poison. Even dissolved in seawater, as its concentrations increase, it becomes lethal to sea life. And if it makes it into the atmosphere in sufficient quantity, it may be a threat to terrestrial life, a result we will revisit in more detail below when we discuss causes of extinctions.

So when and why did the oceans become Canfield oceans? The cause is now known. There is a peculiar type of bacteria that uses sulfur for metabolism. It takes organic carbon and sulfur for substrates and an energy source, giving off a waste product of H_2S. These are known as sulfate-reducing bacteria, a nasty cast of characters of many individual species. They are always down there on the bottom of the oceans, and many have been there for literally eons, but they seem to have bloomed into toxic abundance over much of Precambrian time.

So toxic were the Canfield oceans that they may have inhibited life's first evolution for millions of years during the long ago Precambrian time intervals (tracking back from 600 million years ago to the

77

time of life's origin). There seem to be two reasons behind this thinking. First is the obvious toxicity of the hydrogen sulfide, but just as important may have been the microbe's inhibition of nitrogen formation in compounds useful for plant life. While many kinds of microbes can "fix" nitrogen, an essential element for life, into compounds that are biologically useful, the eukaryotes (or plant life) cannot do this trick and depend upon microbes to do the job for them. Enter the Canfield ocean's gang of sulfur bacteria. Suddenly, little nitrogen becomes unavailable; this kind of bacteria could care less about nitrogen, but it selfishly inhibits other microbes from getting it and supplying it in useful forms to the eukaryotes. A nitrogen-poor ocean would have been an ocean literally in need of fertilizer and not getting it. It would have been just like a soil where all the nitrogen has been leached out somehow—only a small amount of plant life could grow in such a situation. Nasty place, that Canfield ocean.

We are just now learning about the conditions leading to Canfield oceans, and how and why they switch over to either oxygenated or the traditional anoxic ocean (one without the massive quantities of H_2S). And interestingly enough, a Canfield ocean has turned up at several times in the deep past that coincide with catastrophic mass extinctions, as we will see below.

In any event, this seems to be a prime case where life—the sulfur bacteria—made the Earth a worse place for other life, for the evolution of more complex life, and for the evolution of the kind of life (using oxygen) needed to colonize the land areas of our planet. The Canfield oceans thus seem to have held planetary biomass to lower levels than it otherwise would have been.

A Second Snowball, 700 Million Years Ago

The first snowball of 2.3 billion years ago had marked effects on Earth life. Seemingly the most lasting was the lag effect on evolution. After this ancient snowball, life underwent little further evolution toward complexity. Eukaryotic, multicellular life was already present at the time of the first snowball, but it did not diversify for 1.5 billion years. But when new kinds of plants finally did diversify, starting at about 700 million years ago, the Earth entered a new

Snowball Earth episode. This one, if anything, was even more catastrophic because by this time there was more continental surface, which helped bring about much colder global temperatures. This new snowball drastically affected the pathways of nutrient cycling and how it may have delayed the rise of land life by 100 million years. Like the first snowball, this event was also caused by life—and thus was Medean.

The Rise of Animals, the Reduction of Life, 600 Million Years Ago

The end of the snowball of 700 million years ago ushered in a time of change on Earth—great evolutionary change. Soon after, the first animal phyla begin to appear on the planet at a time when the Earth's biomass, recovering from the second Snowball Earth event, had risen to what may have been its highest level, as we shall see in the next chapter. This event also seems to signal the end of the long periods of Canfield oceans, the highly toxic, low oxygen, high hydrogen sulfide ocean bottoms described previously.

The rise of animals that so spectacularly occurred during the 540- to 500-million year-old "Cambrian Explosion" is correctly deemed one of the great evolutionary events ever to have affected the Earth. Clearly the number of species on the planet radically increased. But what of biomass? Just the opposite seems to have occurred. Concomitant with the rise of animals and higher plants, there is a drastic reduction in the number of stromatolites and other evidence of layered bacterial slicks. The evolution of the first herbivores and carnivores among the emergent animals was a major reason for this. The fecal pellets of the newly evolved zooplankton—little animals that feed on plankton and other microorganisms—stuck together and formed slime balls that readily sank down to the abyssal depths, removing organic carbon and nutrients from the sunlit, photic zone and preventing them from being recycled by way of photosynthesis. Thus, we can view the evolution of complex life on our planet as Medean, since it led to a reduction in the total amount of life.

This biomass reduction was probably caused not only by the herbivorous success of the newly evolved animals. There may have been

79

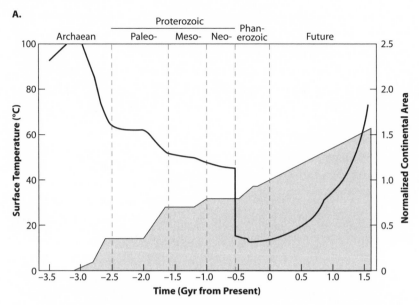

Figure 5.1. Modeled global temperatures, based on model results. Source: Franck et al. (2006).

a significant drop in global temperatures as well—in fact, the largest single drop in planetary temperature in Earth history, if the modeled results suggesting this are correct.

A group of climate modelers led by Siegfried Franck of Potsdam University in Germany have been the leaders in modeling future conditions on Earth, and their work will be discussed in detail in chapter 7. However, they have also modeled past temperatures (fig. 5.1). Their results are startling in one important respect: the models used, based on best estimates of the various parameters used (such as CO_2 content of the atmosphere, as well as continental growth and rates of volcanism and other kinds of tectonism), showed a global surface temperature drop of nearly 40°C, coincident with the appearance of animal life in the Cambrian Explosion. Franck and his colleagues explicitly blame the appearance of animals on this drop. Such a temperature drop would alone have reduced planetary biomass of prokaryotes considerably. Once again, this is a Medean event,

√
Jackson

−Really?

80

and a highly significant one for estimating the frequency of animal life in the cosmos. The Earth survived this radical change because it was were near enough to the Sun and had sufficient CO_2 in the atmosphere to avoid another, and perhaps far more lethal, Snowball Earth episode.

The Phanerozoic Microbial Mass Extinctions, 365–95 Million Years Ago

The discussion of Canfield oceans takes us to mass extinctions occurring since the first evolution of animals. Mass extinctions were short-term events reducing the diversity of life on Earth, and to kill off species requires the wholesale death of the many individuals making up those species. In a mass extinction, there are so many species extinctions that the number of individuals of all kinds of life must be significantly diminished. Mass extinctions play no part in the Gaia hypotheses. However, the Gaians consider them "Gaia neutral," since the groundbreaking discovery in 1980 that the famous dinosaur-killing "KT" mass extinction was caused by an asteroid 10 km in diameter that hit the Earth on what is now the Yucatán Peninsula (leaving Chicxulub crater, 200 km wide). In the scientific aftermath of this paradigm-changing event, it was thought that all but one of the Big Five mass extinctions (Ordovician, Devonian, Permian, Triassic, and Cretaceous) may have been similarly caused, by asteroids or comets, the one exception, the Ordovician event, was ascribed to the effects of a gamma ray burst or nearby supernova explosion; hence, it too had an extraterrestrial origin. For each of the others, as well as for some of the more minor mass extinctions, evidence of impact has been reported in the literature, and in some cases, as for the largest and most catastrophic of all mass extinctions—the Permian event (90 percent of species killed)—the scientific report translated into widespread journalistic stories that brooked no opposing view.

Life on Earth was not capable of predicting asteroid strikes, or preparing for them in any way. Hence these events had a Gaia-neutral standing. However, much recent work on the mass extinctions has refuted the impact hypothesis for all save the original extinction

Gaians
extinction
external
(Problem)
Medean
internal
(but
always
recover =
Gaian?)

event, the KT, which to this day is explained as the result of large-body impact. But what of the others? Their cause seems to have been similar and in fact was life itself—at least, the effects of some species of life, in this case microbes gone wild. As such they become some of the most powerful evidence of all supporting Medea and refuting Gaia.

Only in the past two years have paleontologists and geochemists discovered the existence and the true ferocity of these events, which can be referred to as "greenhouse mass extinctions." They were mass extinction events caused by a poisonous atmospheric state that could return over the next few millennia if carbon dioxide levels reach or exceed 1,000 ppm. The agents of these events live now, deep in the seas of the world, unknowingly awaiting a return to the abundance that they enjoyed for all but several hundred million of the last 2 billion years, a stranglehold that would never have been released but for the slow, 500-million-year-long diminishing of the Earth's atmospheric carbon dioxide as carbon was slowly pulled from the sky and sequestered into coal, oil, and limestone. Now, as a result of our burning, that carbon has come back as carbon dioxide and methane to warm the world, poison the oceans, and potentially return us to a new greenhouse mass extinction, just as it did 490, 360, 251, 201, 190, 135, 100, and perhaps 55 million years ago.

What were these events and how did they come about? They were caused by blooms of sulfur bacteria in the seas, which can only live in the absence of oxygen. This can occur in globally warmed worlds where heat flow from tropics to poles is nil, or in which the thermohaline circulation systems of the deep ocean, which now keep the deep bottoms of the ocean oxygenated, have been shut down. We know, for instance, that some 251 million years ago, CO_2 levels quickly shot up past 1000 ppm due to a gigantic volcanic event that flooded the atmosphere with volcanically produced CO_2. That produced what is now called the Permian extinction, when both ocean and atmosphere warmed over millennial time scales to the point that crocodiles and palms existed at the Arctic Circle, and, more ominously for climate, the heat differential between tropics and the Arctic and Antarctic narrowed to the point that wind and ocean currents

82

petered out to a vast, planetwide calm. As the great current engines slowed and then stopped, normal oceanic mechanisms carrying cold, oxygenated seawater from the surface into the deep sea stopped as well. The deep sea warmed and, like the modern Black Sea, went anoxic, killing all animals in a planet-spanning anoxic zone. Slowly this mass of oxygen-free water moved upward from the deep to the shallow sea, and as it did so an entirely different bacterial flora replaced the normal component of oxygen-loving plankton and microbes. The oceans changed states, and soon the atmosphere did too.

Mass extinctions were short periods of species death, causing entire families and even orders of organisms to go extinct. With the groundbreaking discovery by the Alvarez team from Berkeley that the age of dinosaurs came to a rapid and flaming end 65 million years ago due to the aftereffects of a large asteroid hitting the Earth, a paradigm shift occurred, leading most Earth scientists to conclude that not only the age of dinosaurs, but most, or perhaps all, of the other fifteen or so great mass extinctions of the past (five being particularly catastrophic, with over a 50 percent species kill) were similarly caused by large rocks falling from space. Discoveries were announced in 2001 and in 2002 (both in *Science* magazine to much hoopla) that the largest of the past mass extinctions—the Permian extinction, or Great Dying, and the Triassic mass extinction of 201 million years ago—were also caused by asteroid impacts. Only one, a minor event at the end of the 55-million-year-old Paleocene epoch, seemed different from the impact extinction paradigm, but since it was but a one-time deal, it could be overlooked by the brutish scientific hegemony that impact extinction science had become. In a way, the idea that all mass extinctions were caused by rocks from space seemed a somewhat comforting finding, and *Deep Impact* and *Armageddon*—the two movies of the late 1990s emerging from this extinction research, showing us bravely blasting these dangerous asteroids out of our planet's path—gave further comfort. Our species could surely engineer its way out of any such future event. Then a funny and almost entirely unreported thing happened. The impact evidence of the Permian is now gravely in doubt, and impact at the end of the Triassic was found to have been far too small to kill much beyond the crater's blast area. While "Death by Asteroid" makes a great news headline, the slowly accumulating evidence that quite another cause was involved for all but the dinosaur-killing, KT mass extinction (the sole mass extinction linked to an impact as its major cause) has been entirely overlooked by the press.

But if not impact, what? A new cause had to be found, and the answers came from a new kind of science—from specialists who learned how to extract tiny fragments of cell walls and ancient proteins from ancient rocks. These ancient chemical fragments, called biomarkers, could be related to highly specific biological groups such as specific bacterial (or even animal or plant) orders. By 2005, separate groups working at Curtin University in Australia and the Massachusetts Institute of Technology had found remains (from geographically far-flung, Permian-aged oceanic strata) of a particular sulfur-loving bacterium that could exist only in the sunlit regions of shallow seas that were both anoxic and sulfur rich—hydrogen sulfide rich, to be exact. Such bacteria could exist in the large numbers indicated by the biomarkers only if the oceans were entirely warmed, quiescent, and anoxic from bottom to top. In short order, these same biomarkers were recognized at multiple mass extinction sites spanning 500 million years of time, the time of animals. A new cause of mass extinction had been discerned: global warming, which led to global stagnation and the resultant chain of events—no heat gradient from equator to pole, no currents or wind. Without currents, a warmed ocean loses its oxygen, from the bottom up. When that happened, a group of microbes now found only in very small numbers in the few low-oxygen pockets of our seas bloomed into abundance. One of these is a group of sulfur-utilizing bacteria that produce the highly toxic gas hydrogen sulfide as a byproduct of their metabolism. Hydrogen sulfide kills animals even at low concentrations, and the rock record now shows repeated episodes when large volumes of this gas came out of solution from the sea, entered the atmosphere, and, amid the high heat of the air, managed to gruesomely kill off most land life, especially plant life. This happened at least eight times that we know of, and more such events are being discovered every few months (the "army" of geobiologists is small and funding miniscule).

These changes do not happen overnight: the change from our current swept world to one where there are no oceanic currents—the necessary forerunner to the anoxic ocean that the sulfur bacteria need—will take place at millennial scales once, and if we exceed atmospheric CO_2 levels of about 1,000 ppm. But at that level of carbon dioxide, the rock record tells us that there cannot be ice caps. If the ice caps melt, the chain of events leading to another greenhouse extinction will be set in motion. The most extreme of these past events killed off 90 percent of all species—and to do that, all individual life well in excess of 99 percent would have been wiped out.

The numerous greenhouse extinctions of the past were triggered by excess volcanic activity in the form of vast lava fields known as flood basalts. This time the proximal cause will be different, but carbon dioxide is carbon dioxide, be it from a volcano or a Volvo. But in either case, the role of life (microbes) in causing the actual "kill" mechanism is strongly Medean, not Gaian. Once again, these events are evidence that the Gaia hypothesis should be discarded.

Rapid Global Temperature Changes Due to Colonization of
Land by Plants, 400–250 Million Years Ago

The time interval from 400 to 250 million years ago seems to have been one of great temperature swings. Beginning some 400 million years ago, the world was warm; from 350 to 300 there were enormous glacial intervals, and then it became warm and desertlike at the end of the Permian. Such disturbances do not foster increased diversity. That they were caused by life indicates that they were Medean.

The rapid changes of temperature through this time may be attributed in part to the evolution of land plants, for their presence has had an enormous influence on the Earth's climate and mean temperature. Because of their need to anchor in substrate and adsorb water and nutrients through rooting systems, vascular plants have helped create soils and have markedly affected the weathering rate of rocks, which thus affects the amount of carbon dioxide in the atmosphere.

Vascular plants not only affect the rates of weathering through the mechanical breakup of the material they root in. They also have a marked effect on the chemical weathering of rocks. Plant rootlets (plus the symbiotic microflora such as bacteria and fungi that they contain) have a very high surface area, across which they secrete various organic acids that attack the substrata minerals in order to harvest nutrients such as phosphates, nitrates, and various elements necessary for growth. In addition, the plants, after dying, produce organic litter, which decomposes to organic acids, and H_2CO_3, which provides additional acid for the chemical breakdown of minerals. The roots of plants also markedly affect the water retention of soil, for the roots anchor clay-rich soils against erosion. The retention of water in this type of soil thus maintains a liquid environment around mineral grains, accelerating the dissolution of mineral material. All of these factors combine to accelerate the weathering of rocks, which ultimately drives down atmospheric carbon dioxide.

Prior to the evolution of vascular plants in the Devonian period, some 400 million years ago, there must have been a very different rate of chemical weathering on land, and hence far higher atmospheric CO_2 and planetary temperature. Although some scientists have suggested that there were sufficient fungi and algae on the Earth's surface prior to this time to affect chemical weathering rates, most agree that it was the evolution of root systems that was the most significant change. The actual enhancement of weathering rates has now been studied experimentally. It appears that the presence of vascular plants accelerates weathering by about four to ten times.

The evolution of vascular plants had a second effect on decreasing atmospheric CO_2, and thus temperature rates. As plants die, they fall and become buried in sediment. Although much of this material decomposes, significant amounts, especially those parts of the plant composed of the woody material lignin, are resistant to microbial decomposition, especially if the material is rapidly buried. The burial of organic material (quite often resulting in the formation of coal deposits, which are geologically stable for millions to hundred of millions of years) removes carbon from the atmosphere and transfers it into stable, stratigraphically bound reserves. The increased rates of carbon burial, starting in the 400-million-year-old Devonian period (the time when land plants became abundant on the continents) led to a dramatic decrease in atmospheric CO_2. Vascular plants thus provide two main ways of lowering atmospheric CO_2—by increasing the rate of silicate rock weathering, and by transforming inorganic carbon to organic carbon and then removing it from the "open" air-water systems into relatively closed sedimentary deposits.

The climatological implications of this major revolution—the evolution of land plants—led to a twentyfold decrease in atmospheric CO_2 over a 100-million-year period. A major consequence of this was a dramatic cooling of planet Earth, and the creation of a major glaciation episode affecting large parts of the planet, some 350 million years ago. Since that time, the overall amount of inorganic carbon cycling through the system has been decreasing enough to cause a long-term temperature drop.

Devonian Eutrophication Events, 360 Million Years Ago

As recounted in chapter 4, only recently have students of the deep past recognized what are hypothesized to have been oceanic eutrophication events. These are times when short-term blooms in plankton, themselves the product of an anomalous increase in surface nutrients, which trigger the bloom, die off, sink to the bottom, and ultimately are scavenged by microbes with the effect of using up all dissolved oxygen at the base of the ocean. An example of this is the Devonian mass extinction, another of the Big Five mass extinctions, which happened about 360 million years ago. Various scientists and researchers contributed to this discovery, notably Thomas Algeo and Brad Sageman, two geologists specializing in research on such events.

While superficially resembling aspects of the greenhouse extinctions detailed above, especially since both involve oxygen-free bottom water, there are, nevertheless, important differences between eutrophication events and greenhouse extinctions. The most important might be in the fate of the organic carbon.

During photosynthesis, carbon, nitrogen and phosphorous are incorporated in algal organic matter. When oxygen levels vary during a eutrophication event, bacteria in the lower water column and sediments tend to release more nitrogen and phosphorous back into the sea as they decompose dead organic matter. Consequently, a surplus of carbon gets buried, and this changes the very nature of sediment—from limestone-dominated (with most of the lime coming from the skeletons of organisms, including reef formers such as corals and sponges) to black shale, which harbors a completely different suite of organisms. And as decomposition increases, the bottom waters lose whatever oxygen they had. Another important difference from the greenhouse extinctions is that these eutrophication events are posited to occur during times of low, not hot, global temperatures.

The net results of the Devonian extinction are stunning. In my own field area, the Canning Basin of Western Australia, one of the largest and best-preserved ancient coral reefs in the world changes from animal-dominated to microbe-dominated as a result of the extinction. There is evidence for low-oxygen water and the deposition of dark shales amid the shallow-water limestones, indicating that the low-oxygen water came right up to the surface of the sea, killing off the shallow-water reefs.

The Devonian event shows that these eutrophication episodes decimated biomass as well as diversity. They are certainly Medean.

The KT Extinction, 65 Million Years Ago

It was noted above that the asteroid-instigated KT extinction of 65 million years ago was thought to be "Gaia neutral"; because it came about due to an extraterrestrial impact, it had nothing to do with life. While it is true that the impact was extraterrestrial, it could be argued that the effects of life magnified the extent of the extinction.

By the end of the Cretaceous there were planet-spanning forests. One effect of the impact was the ignition of continent-spanning forest fires. This produced an enormous amount of ash and dark carbon soot that filled the atmosphere. This soot, a product of life, caused global cooling for some months after the impact, which seems to have played a significant role in the kill mechanism. Again, a Medean effect.

The Pleistocene Ice Ages

If we were to travel back in time, some eighteen thousand years ago, we would find a world quite different from that with which we are familiar today, and this view may serve as our guide for what the next ice age might do once again to the planet. Many parts of the

Northern Hemisphere that are now densely inhabited were (and will be again) covered in ice or permafrost, for glaciers reached as far south as what are now New York City and much of central Europe, and gigantic regions to the south of these areas were permanently frozen. The Atlantic Ocean was clogged with icebergs. In many parts of North America and Europe, the glaciers were as much as 3 km thick, and the regions just south of the ice were as inhospitable as the ice-covered regions themselves due to enormous winds generated along the fronts of the glaciers. Wind speeds as high as 300 km per hour, greater than almost any hurricane today, were commonplace along the edges of the stupendous continental glaciers of the time. Such huge winds would have pushed mountains of dust and sand through the air to pile and unpile along the glacial fronts. A treeless, tundralike landscape extended for hundreds of miles south of the glaciers themselves. Farther south great deserts existed because the world was so much dryer. Even the tropics were disrupted. The Amazonian rain forest disappeared, replaced by pockets of lush vegetation surrounded by oceans of savanna.

In spite of the enormous changes that the ice ages visited upon the Earth, prior to the nineteenth century their existence was unknown to science. Naturalist Louis Agassiz was among the first to realize that many of the Earth's topographic features could only be explained as the product of continental ice cover of recent antiquity. The presence of boulders and ridges crossing flat lands, enormous isolated boulders far from any rocky outcrop, ancient channels of enormous rivers not now present, and deeply cut lake basins could only be explained if much of northern Europe and North America had been covered by ice.

An even greater surprise was the more recent discovery that this continental ice cover had not happened just once, but repeatedly. The use of oxygen isotopes from marine plankton records, along with the advent of reliable radiometric dating in the 1950s and 1960s, filled in the dates of the glaciation. These had not been times from great antiquity, as measured in tens of million of years ago, but far more recent events dating from the past 2 million to only a few tens of thousands of years ago.

Why did the Earth undergo these radical and catastrophic cooling events? These ice ages, the first of the last 300 million years, took place because of the long-term decrease in atmospheric carbon dioxide, and, as we have seen, *that reduction has been caused by life.* During these Pleistocene ice ages, planetary productivity and biomass plummeted.

• • •

These events of the past certainly were a rogue's gallery of planetary antibiotics. But there is at least one more to add to the list (for there will surely be more ancient events that eventually get blamed on life itself), and that anti-Gaian agent is humanity. So intense is our effect on reducing the biomass of our own planet that the whole next chapter will be devoted to the subject.

And why did they END?

6

HUMANS AS MEDEANS

We do not have to wait for the slow meandering of evolution to adapt us to the altered climate and atmospheric chemistry our guild is now creating
—Tyler Volk, *Gaia's Body*, 1998

One has only to be an aficionado of futuristic cinema to get a sense of how really BAD we humans are. The entire post-apocalyptic genre—the high-water mark of Blade Runner, such oldies as Soylent Green, THX 1138, the Mad Max epics, A Boy and His Dog, the Planet of the Apes old and new—points to a future that really looks not only dreadful, but dead, in most cases. From a productivity point of view, those futures look both bleak (for us humans) and positively post–mass extinction in terms of biotic biomass. Be it polluted cities (Blade Runner, Soylent Green) or desert landscapes (the Mad Maxes and so many others), we see a vision of little life other than humanity. This is completely consistent with a Medean result: that humanity would reproduce to numbers such that it would cause a reduction not only of biodiversity, but of biomass. Planets do not care about anything, and if an objective measure of biomass is taken, it matters not if there is one species or millions making up the total—it is the bottom line that counts. As we shall see in the next chapter, the high point of the Earth's biomass occurred a billion years ago or so—long before animals, complex planets, and the high biodiversity of the post–Cambrian Explosion world.

HUMANS AS BEHAVIORAL PROKARYOTES

The most fundamental division in Earth life is between the prokaryotes and the eukaryotes. The latter are composed of two distinct taxonomic assemblages termed domains – these are the bacteria and archaea, morphologically similar microbes that are genetically so distinct that they merit status in these high taxon categories. The second great category, the eukaryotes, differ from the prokaryotes by being larger and, more important, by containing their genetic material in an enclosed nucleus within the cell itself. The eukaryotes also contain a number of smaller membrane-enclosed organelles, such as mitochondria, ribosomes, and (in plants) chloroplasts, which have been interpreted as having a separate prokaryotic origin but were subsequently subsumed genetically by the larger cells through initial symbiosis, and later complete genetic aggrandizement.

It is the eukaryotes that best evolved into multicellular organisms. While some prokaryotes also evolved multicellular, among them there is nowhere near the size or internal organization found in eukaryotes.

Of the many fundamental differences between the prokaryotes and eukaryotes, there is one that could be called "behavioral" in a rough sense. Not by the individuals, but in an evolutionary way. When confronted by environmental challenges that are lethal, prokaryotes respond by trying to change their environment, as well as themselves. Because so many prokaryotes are so successful at producing a variety of chemicals that they extrude from their bodies, they often overcome the challenges presented by changing the environment more than they change themselves. For example, if confronted by a medium more acidic than is comfortable, Darwinian forces cause the evolution of more acid-tolerating microbes. But at the same time, the population of microbes might as well secrete chemicals making the solution they rest in more basic, and thus lowering the acidity.

Eukaryotes respond quite differently. In the face of challenges, they change their morphologies. In the case above, instead of trying to reduce acidity, they might evolve the means to escape the

environment—or produce a cell wall more protective when surrounded by an acid medium. It is through morphological change
that eukaryotes evolved. The prokaryotes do not do this, they remain
one of their three shapes—rods, balls, or spirals—no matter the challenge. But internally, the chemical systems evolve quickly.

There is another way the prokaryotes and eukaryotes differ. In terms
of major environmental changes affecting the planet, the prokaryotes
in most cases win hands down. It was microbes that changed the
early Earth atmosphere from one of nitrogen and carbon dioxide to
an enriched methane atmosphere, and then changed it again to one
that was oxygen-rich. The subsequent action of microbes has been
instrumental in all the geochemical cycles, and the perturbations in
greenhouse gases, strongly affected by microbial action, have caused
great global temperature swings. In contrast, eukaryotes have had a
less marked effect on the planet. The most important of these effects
have been produced by plants, especially rooted plants, which change
weathering rates and thus affect the silicon-carbonate thermostat of
the Earth.

This divide in evolutionary style has remained consistent for a very
long time. For the first time, however, one species of eukaryotes has
begun to act more like microbes: us. Our first examples of this are
many, but mostly crude. In the face of cold, we do not make the
world warmer, as microbes would (not yet anyway, though we could
now); we put on clothes. Dealing with heat is harder and more technologically challenging, but we are tackling even that through air
conditioning, by causing the actual temperature to drop around us.

Increasingly, as changing conditions around us challenge us, it is
through this prokaryotic means of survival that we will persevere.
Already many are considering colonies on Mars—and how to "terraform" that planet to eventually allow humans the ability to live
there. This is all very prokaryotic.

Yet while this way of engineering mimics the positive ways of im-
proving environments, unfortunately we also mimic the prokaryotes
in the sense that, per capita, we are changing the environment around
us more than are any other eukaryotes. The byproducts of many
microbes are extremely toxic to other organisms, and even to the

microbes themselves. The production of methane, or alcohol, or hydrogen sulfide, or even heat by runaway population growth in microbes can kill not only other organisms, but also the microbes themselves. The vast array of toxic waste produced by humanity, from too much carbon dioxide to too much plutonium waste, is unprecedented even by most microbial actions of the past.

Let us look at other Medean traits manifested by humanity.

HUMAN POPULATION

More than 200 years ago, the British scientist Thomas Malthus described the single most intractable problem with human population growth. While our population numbers increase exponentially, human food supply tends to increase on a linear scale as more land is devoted to agriculture. The inescapable conclusion is that human population will thus tend to outgrow its food supply. In related fashion, human population is likely to outstrip its supply of fresh, untainted, and unpolluted water.

Ten thousand years ago there may have been at most 2 to 3 million humans scattered around the globe. There were no cities, no great population centers; humans were rare beasts, living in clans, nomadic groups, or, at best, settlements with little lasting construction. There were fewer people on the entire globe than are now found in virtually any large America city. Two thousand years ago, the number had swelled almost a hundred fold, to 130 million or perhaps as many as 200 million people. The one billion mark was reached in 1800; there were 2 billion people in 1930, 2.5 billion in 1950, 5.7 billion in 1995, and approximately 6.5 billion in 2000. At this rate of growth, the human population is expected to exceed 10 billion by 2050 to 2100, assuming an annual increase of 1.6 percent. While this rate is somewhat reduced from the 2.1 percent characterizing the 1960s, it remains a staggering figure. In 1992 the United Nations published a landmark study calculating potential human population trends, arriving at several estimates. By 2150 the human population could reach 12 billion if human fertility figures fall from present-day levels of 3.3 children per woman to 2.5 children. If, however, faster-growing regions of the world continue to increase in population and maintain their current fertility levels, average fertility worldwide will increase to 5.7 children per woman, and the human population could exceed 100 billion people sometime between 2100 and 2200. This latter figure seems beyond the carrying capacity of the planet. Officially, the United Nations uses three estimates for the year 2150: a low estimate of 4.3 billion, a medium estimate of 11.5 billion, and a high estimate of 28 billion.

Predicting future population numbers is a difficult endeavor because of the many variables involved. The definitive work in recent times is Joel Cohen's 1995 book *How Many People Can the Earth Support?* (pp. 367, 369). Cohen's conclusions are stark:

> [T]the possibility must be considered seriously that the number of people on the Earth has reached, or will reach within half a century, the maximum number the Earth can support in modes of life that we and our children and their children will choose to want.... Efforts to satisfy human wants require time, and the time required may be longer than the finite time available to individuals. There is a race between the complexity of the problems that are generated by increasing human numbers and the ability of humans to comprehend and solve those problems.

Humans depend for food largely upon a narrow range of staple crops that are dominated by grasses and several species of livestock, a system that was established and has remained essentially similar for 300 to 400 human generations.

HUMAN-INDUCED BIOMASS REDUCTION

Forests have been a part of this planet for more than 300 million years, and although the nature of species has changed over that long period of years, the nature of the forests has changed little. Humankind, over the brief moment of history that we have been on this planet, has treated the forests as if they were infinite, as indeed they must have seemed for much of our existence on this planet.

The forests are the great arks of species on this planet. Although the land surface of our globe is only one third that of the oceans, it appears that 80–90 percent of the total biodiversity of the planet is found on land, and most of that is found in tropical forests. As we destroy these forests, we destroy species. In the late 1990s botanist Peter Raven estimated that 6–7 million species of organisms live in the tropical rain forests, and that only about 5 percent of these are known to science. Because we have such a poor understanding of how many species really exist, it is next to impossible to derive hard figures about how many have gone extinct in the past century, or in the past decade, next decade, or next century.

There appear to be several driving forces causing a reduction of biodiversity—a *destruction* of biodiversity, to be less delicate. The huge run-up of human population completely changes the nature of problems on Earth, and it goes without saying that pressures on our planet's environment are completely unlike those it has experienced at any time in its past.

It is not only the number of people on Earth that has changed, but where they are found. In 1950 about one-third of the human population lived in what we euphemistically call "industrialized," or "developed" counties. In 1995 that number had dropped to about one-fifth, and it should drop to about one-sixth by 2020. The population that the United States represents is about 4.5 percent of total human population. Americans, however, are well represented—if not in numbers, then at least in the effect we have on the globe. For instance, Raven estimated that the humans living in the United States produced 25–30 percent of total world pollution. At present the United States controls 20 percent of the total global economy. Of the 3,000 culturally and linguistically distinct groups of humans found on Earth, the population of humans calling themselves "Americans" is the wealthiest, and the wealthiest in the history of the planet. One consequence of this is that we consume more of the resources produced by the Earth than any other country. We also use more energy per capita than any other county, by far—and why not? Since 1945 (and until the big price increase of 2008) the cost of gasoline in the United States dropped by 33 percent in adjusted costs.

Much of industrialization has been at the expense of the forests. Forest conversion—a conversion that changes forests first to fields and then usually to overgrazed, eroded, and infertile land within a generation—is perhaps the most direct cause of biodiversity loss. According to my colleague David Montgomery, it appears that 25 percent of the world's topsoil has been lost since 1945; by lost Montgomery means that it has been stripped from the surface and redeposited either in the seas or in deserts. According to biologists such as Jared Diamond, about one-third of the world's forest disappeared in the same interval, while 40 percent of the total plant production (measured by photosynthetic productivity) is now used by humans in some way—such as food, timber, or grazing land. Such figures should cause alarm, and in some cases they are meant to. These figures might be too high (or perhaps too low), so it is useful to have skeptics demand better accounting of such losses, as in the 2001 book by Bjorn Lomborg, *The Skeptical Environmentalist*. Nevertheless, there have been losses, and they certainly have been significant, as anyone who has ever flown over the formerly forested western states of the United States can attest.

The Medean aspect of this is that the removal of forests—to be replaced first by fields, and then, all too often, by rock, once soil cover is eroded away—produces a net reduction of planetary biomass. Tropical rain forests are probably regions with the highest overall biomass of the planet per unit area; they are certainly the highest in productivity. Since the vast majority of the oceans, though volumetrically so much greater than the land area, are essentially empty of life, it is on the continents that most biomass resides.

In summary, humans manifest Medean characteristics. (And, in a sense, how could they not? Medea herself was a human, Jason's wife, and murderer of their own children.) But on a more serious note, we humans are indeed reducing planetary biomass in ways completely consistent with the Medea hypothesis—and in ways inconsistent with Gaia.

7

BIOMASS THROUGH TIME
AS A TEST

Strange is our situation here on Earth.

—Albert Einstein

The Medea hypothesis supports the view that life decreases the prospects for more life. Therefore it can be shown that biomass will eventually decrease through time and in fact is doing so now, as we will see in this chapter. Here we will look at two different ways of judging planetary biotic "success"—through diversity and biomass through time. We will begin with biodiversity. Has the change in species through time followed patterns predicted by the Medea hypothesis, or some other pathway?

DOES THE HISTORY OF BIODIVERSITY SUPPORT ONE HYPOTHESIS OVER THE OTHER?

Animal and Higher Plant Diversity

The history of biodiversity—the assembly and measurement of diversity and biomass through time—was first considered in the work of John Phillips, who is credited with subdividing the geological time scale through the introduction of the concepts of the Paleozoic, Mesozoic, and Cenozoic eras. Phillips, who published his monumental work in 1860, recognized that major mass extinctions in the past could be used to subdivide geological time, since the aftermath of each such event resulted in the appearance of a new fauna as recognized in the fossil record. But Phillips did far more than recognize the importance of past mass extinctions and define new geological time terms: he proposed that diversity in the past was far lower than in the modern day, and that the rise of biodiversity has been one of wholesale increases in the number of species, except during and immediately after the mass extinctions. His scheme recognized that mass extinctions slowed down diversity, but only temporarily.

Phillips's view of the history of diversity was completely novel. Yet a century passed before the topic was again given scientific attention. In the late 1960s paleontologists Normal Newell and James Valentine again considered the problem of exactly when, and at what rate, the world became populated with species of animals and plants. Both wondered if the real pattern of diversification was a rapid increase in species following the so-called Cambrian Explosion of about 540–520 million years ago, followed by an approximate steady state. Their arguments rested on the importance of so-called preservation biases. Perhaps the pattern of increasing diversification through time seen by Phillips was, in reality, simply the record of preservation through time, rather than the real evolutionary pattern of diversification. According to this argument, the change of species is reduced in ever-older rocks, so that sampling bias is the real agent producing the so-called diversification seen by Phillips. This view was soon after echoed by paleontologist David Raup; in a series of papers, he forcefully argued that there are strong biases against older species being discovered and named by scientists, since older rocks experience more alteration through recrystallization, burial, and metamorphism. In this manner, entire regions or biogeographic provinces have been lost to time (therefore reducing the record for older rocks); there is simply more rock of younger age to be searched.

The argument as to whether diversity has shown a rapid increase through time, or whether it achieved a high level early on and has stayed approximately steady ever since, dominated paleontological research for much of the latter part of the twentieth century. In the 1970s massive data sets derived from published records of fossil appearances and disappearances began to be assembled by the late John Sepkoski of the University of Chicago (and his colleagues and students). These data, compiling the record of marine invertebrates in the sea, as well as other data sets for both terrestrial plants and for vertebrate animals, seemed to vindicate Phillips's early view. In particular, the curves discovered by Sepkoski showed a quite striking record, with three main pulses of diversification carried out by different assemblages of organisms. The first was seen in the Cambrian (the so-called Cambrian fauna were composed of trilobites, brachiopods, and other archaic invertebrates), and this was followed by a second in the Ordovician. The Ordovician led to an approximate steady state throughout the rest of the Paleozoic (the Paleozoic fauna were composed of reef-building corals, articulate brachiopods, cephalopods, and archaic echinoderms) but culminated in a rapid increase beginning in the Mesozoic. Diversification then quickly accelerated in the Cenozoic to produce the high levels of diversity seen in the world today. The evolution of the modern fauna happened during this time—gastropods and bivalve mollusks, most vertebrates, and echinoids, among other groups.

The net view of biodiversity over the last 500 million years was the same as that of John Phillips in 1860—there are more species on the planet than at any time in the past. Even more comforting, the trajectory of biodiversity seemed to show that the engine of diversification—the processes producing new species—was in high gear, suggesting that in the future the planet would continue to have ever more species. While not at the time viewed in any sort of astrobiological context, these findings certainly do not suggest that the Earth is in any sort of planetary old age. All in all, the 130-year belief, from the time and work of John Phillips to that of John Sepkoski—that there are more species now than at any time in the past—remained a comforting view. This long-held scientific belief suggested to many that we are in the best of biological times (at least in terms of global biodiversity), and that there is every reason to believe that better times, an even more diverse and productive world, still lie ahead.

While Sepkoski's work seemed to show a world where runaway diversification is a hallmark of the late Mesozoic into the modern day, worries about the very real sampling biases described by earlier workers persisted, and a series of independent tests of diversity were conducted. Of most concern was a phenomenon dubbed "the pull of the recent"—that the methodology used by Sepkoski would under-

But we don't know # spp today to nearest order of magnitude - so debate is uninformed

count diversity in the deep past, making it look like there were ever more species in more recent times. Because of this very real concern, new tests were devised to examine biological diversity through time. In the early part of the new millennium, the issue was reexamined by a large team headed by Charles Marshall and John Alroy. This team assembled a more comprehensive database based on actual museum collections, rather than Sepkoski's method of simply tabulating the number of species recorded in the scientific literature for given intervals of past geologic time. To virtually everyone's surprise, the first results of this effort were radically different from the long-accepted view.

The analyses of the Marshall–Alroy group found that *diversity in the Paleozoic was about the same as in the mid-Cenozoic*. The dramatic run-up of species that had so long been postulated for diversity through time was not evident in this new study. The implications are stark: we may have reached a steady state of diversity hundreds of millions of years ago. As we will see, this new finding might be in accord with important new work by astrobiologist S. Franck and his colleagues on the abundance (not diversity) of life through time.

It may be that diversity peaked early in the history of animals, in contrast to all views since the time of Phillips (prior to the Marshall–Alroy work) that it has remained in an approximate steady state since, or perhaps may already be in decline. While many new innovations, such as the adaptation allowing the evolution of land plants and animals, surely caused there to be many new species added to the planet's biodiversity total, it may be that by late in Paleozoic time the number of species on the planet was approximately constant. The implication of this for our thesis is important: perhaps our planet, rather than still growing in biodiversity totals, has already peaked and is sliding back into lesser numbers, just as the various models by the Franck group and others suggest that global productivity may have peaked hundreds of millions of years ago; we have already seen our best days. This finding is consistent with predictions from the Medea hypothesis. It is inconsistent with predictions of the Gaia hypothesis and is another reason why I advocate rejection of that hypothesis.

101

Microbial Diversity

There is one group that is virtually opaque to the fossil record and methods described above, because its members rarely leave fossils that can be identified as distinct species—that group is the microbes, such as bacteria and archeans. Have the number of microbial species in these two groups shown the same trend as the organisms that leave behind body fossils? Here we have no data. Yet we do know that, in the past, there may have been many more kinds of microbes than in the present, judging from the fact that prior to animals the most abundant life on Earth was microbial—and from that, most microbiologists have concluded that in the past there may have been far more microbial species as well as abundance than today, so many as to dwarf the puny trends of the giants of the world, the animals and plants.

PLANETARY BIOMASS— ELUSIVE EVIDENCE

All of the work to date equates diversity with some kind of "success." However, there is a second measure of life that might be far more important in objectively measuring "success" than diversity, and that is abundance. As we saw in chapter 2, biomass and productivity are two ways of measuring the amount of life on the planet—the first is the total weight of living material, and the second is the rate at which inorganic carbon (in the form of CO_2, an oxidized carbon compound) is transformed into organic carbon (reduced, longer chain molecules with abundant carbon). So, what has been the history of biomass over time, and does that metric support or negate the Medea hypothesis?

There is no direct way of measuring past biomass. We are left with modeling. While the old Russian adage reminds us that there were only two kinds of really evil humans—communists and statisticians (and replace modeler for statistician here)—there are reasons for having some confidence in a series of models produced by separate groups, each attempting to answer the question of what biomass was in the past, and what it might be in the future. It turns out that the same models can be used for each. Let us look in some detail at this line of work.

The biomass on the planet should be related to several factors, any one of which can be limiting. The first is energy, the second nutrients, and the third temperature. Energy ultimately comes from the Sun. Since the formation of the Earth, the Sun has increased energy output by about a third. We would thus suspect that in a perfect world, where neither temperature nor nutrients are limiting, the biomass of the Earth should roughly increase through time. But we are not in a perfect world, and this is where Darwinian evolution comes in. The first life was surely less efficient at extracting energy than life is today. Even at the microbial level, there are a number of ways to harness energy from the Sun, but all involve oxygen reduction gradients—in other words, energy acquisition involves chemical changes of taking a compound, oxidizing it, and extracting the energy from that change. Early life may have had any number of energy pathways, but all would have been anaerobic; the metabolism of these organisms does not use oxygen, and in many cases it is even poisoned by the presence of even small amounts of oxygen either dissolved in water or found in the atmosphere. One such way is methanotrophy—taking the compound methane, which was probably one of the major atmospheric constituents of the earliest Earth atmosphere (according to new work by Jim Kasting and David Catling) and using that energy to run the machinery of life through methane-driven metabolism. Another process is fermentation, which involves the formation of alcohol as a byproduct and again is in the absence of oxygen. But it was with the evolution of oxidative mechanisms, specifically following the evolution of oxygen-releasing photosynthesis, that the greatest energy acquisition came about. David Catling has suggested that life throughout the universe should follow a similar path, with an end state of oxidative metabolism, simply because the physics and chemistry of the universe could produce no better way for life to get energy than oxygen-dependent systems—no other process has such a large energy yield.

How might biomass have been affected, then, as this evolutionary process took place? Assuming that life would increase to the point that it would use all resources (one of the properties of Darwinian life, as we have seen), the successive increase in energy acquisition through evolving metabolic pathways should yield ever-higher planetary biomass.

[handwritten margin notes:] No — extra caveats

Evol as progressive toward > efficiency but need only be adequate

A second requirement for biomass is nutrients. We animals ultimately derive our energy from plants. Whether it is by eating other animals that have eaten plants, or by eating plants themselves, we are ultimately plant carnivores. But what of the plants themselves? They also need three elements that can be limiting; carbon, nitrogen, and phosphorus. Their carbon comes from the atmosphere, in the form of carbon dioxide. As we know from any garden or the fate of phytoplankton in the sea, nitrogen and phosphorus, the so-called fertilizers of plants, also have a major effect. Yet of these three, it is carbon that holds the fate of life on Earth—and the most important variable in the "carbon cycle" is carbon dioxide.

THE CARBON DIOXIDE HISTORY OF THE ANCIENT EARTH

In an earlier chapter we profiled the history of CO_2 over the last 550 million years, the time of animals, showing that, although there have been a series of fluctuations, the overall trend has been of decreasing levels of CO_2 through time. But to understand how the amount of biomass has changed even further back in time, we need to find values of CO_2 on the older Earth, during the long period before animals and land plants. Such an estimate has recently been made by the ubiquitous Siegfried Franck. That curve, with error bars, has a lower resolution than does our curve for the time of animals; for instance, the short-term rises in CO_2 at the end of the Permian and Triassic (as well as other times) are not picked up by this method. This long-term curve is shown in figure 7.1.

The striking aspect of this figure is the long-term trend of decreasing CO_2, with a drop of as much as five orders of magnitude. Current levels are at 380 parts per million, and thus the earliest Earth may have had 10,000 times more CO_2. Unlike today, that would mean that a significant percentage—perhaps a third—of the ancient atmosphere would have been CO_2, not the trace it makes up today.

The Temperature History of the Earth

Another major consideration for determining the viability of any organism is temperature. The chemical reactions that are required of

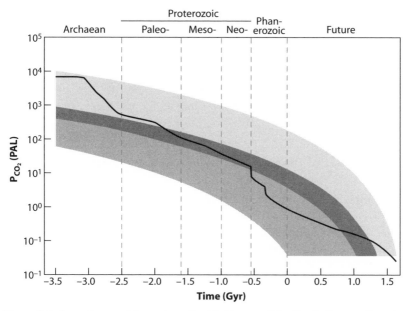

Figure 7.1. An estimate of the carbon dioxide history of the Earth. Source: Franck et al. (2006).

Earth life are also highly temperature sensitive. Because life needs liquid water, this leads to the rather narrow temperature range within which Earth life can live. That range is from slightly below 0°C to slightly above. But biomass at these extremes is low. The 20–70°C range leads to far higher biomass.

How wide have the fluctuations in global temperature been, and when did the global thermostat first kick in? The history of our planet's temperature is not easily studied. There are no direct "paleothermometers" that give some mean global temperature at any given time. While there are a few ways of measuring ancient temperatures from extinct organisms, such as studying the ratio of isotopes of oxygen as recorded in sedimentary rocks, these records are for individual locales rather than the planet as a whole and are applicable mainly over the last 100 million years of Earth history. Inferring ancient paleotemperature thus relies on indirect evidence from the geological and paleontological record. Among such clues is the presence of specific sedimentary rocks indicative of ancient climate (for example, sedimentary rocks known as evaporates, such as ancient salt deposits, are indicative of heat, while glacial deposits tell us of ancient cold). Fossils are also of great importance, for specific types of organisms are often useful in interpreting ancient climates. Fossil soil types are similarly useful, for soils are highly climate sensitive.

Using such methods, paleoclimatologists have arrived at an accepted record of the last 542 million years, the time of abundantly skeletonized fossils. This interval of time, which is composed of the Paleozoic, Mesozoic, and Cenozoic eras, has been shown to be both warmer and colder than the present-day mean temperature of about 15°C. But the temperature variation has not been large—only as much as 10°C either hotter or colder at any time during this long roll of history. We thus may have had an Earth as hot as 25°C, and as cold as 5°C. Neither of these extremes would have imperiled the continued existence of animals and plants on the planet.

While a half billion years is indeed an enormous interval of time, in reality it represents only about the last 10 percent of Earth history. For the other 90 percent of time, we must rely on inferences.

The first review of the temperature record of our planet prior to 500 million years ago was published in the early 1980s. This record suggested a relative rapid cooling of global temperature from about 80°C or higher at 3.8 billion years ago to approximately 40°C or less at 3 billion years ago, to less than 20C by 2 billion years ago, with global temperature never exceeding 30°C subsequent to that time. This interpretation suggested that temperature had little to do with the evolution of life, subsequent to about 3 billion years ago. However, this view is no longer universally accepted. Oxygen isotope records derived from a series of pristine chert rocks that had been deposited well before the advent of skeletonized animals gives a very different story, with global temperatures of above 70°C at 3 billion years ago, 60°C at 2 billion years ago, and about 40°C as late as 1 billion years ago to as recent as 0.5 billion years ago. This new record is shown in figure 7.2, a graph from the Franck group.

Seems
Gaian

√

Steve
>.
(Potsdam
Univ.)

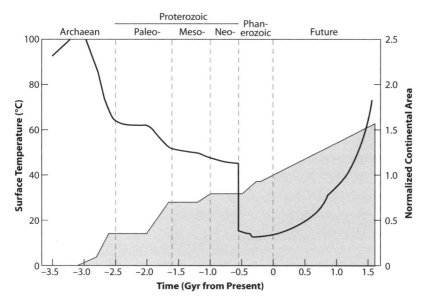

Figure 7.2. Temperature of the Earth, past, present, and future. This figure also appeared in chapter 5. It is shown again here because of its importance for future estimates of biomass. Source: Franck et al. (2006).

BIOMASS ESTIMATES

With CO_2 and temperature estimates (along with other parameters such as rate of continental growth) now firmly in hand, it is possible to model past (and future) biomass. In this chapter I stress past biomass, whereas the next chapter looks at future biomass. (The figures here, however, show both.). This kind of modeling has been done for the Earth by various groups of scientists. But by far the most sophisticated results have come from the Potsdam University team headed by S. Franck, whose group has now done several generations of this kind of model. Their earlier results, from models published in 2000 and 2002, are shown in figure 7.3.

The differing results come from using different starting conditions of temperature and carbon dioxide, as well as estimates of continental size and its growth through time. What is striking and similar,

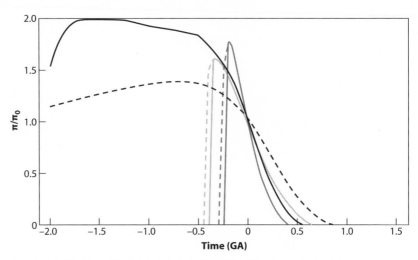

Figure 7.3. Various estimates of global biomass (vertical axis 0 plotted against time, where the past is on the left [minus values], the present 0, and the future positive numbers are on the right). Source: Franck et al. (2000, 2002).

however, is the surprising result that biomass may well have peaked in the distant past—perhaps a half billion years ago, or even earlier, according to some of the models. Only one of the estimates puts maximum biomass anywhere near our time, and even this seems to show a peak some hundreds of millions of years ago. The second result, uniting all the curves, is that biomass is currently falling and in every case would reach 0 between 0.5 and 1 billion years from now. Two things—dropping carbon dioxide levels, and the appearance of animals themselves—are supposed to have caused this, and if so, this is a very Medean result. But the most interesting aspect might be that the graph shown in figure 7.3 seems to suggest that diversity and productivity, or biomass, are roughly independent.

CAVEATS: POTENTIAL PROBLEMS WITH THE MODELS

The models inferring temperature and CO_2 levels are probably giving us at least a reasonable estimate of trends. But for biomass they may be off—way off, in fact. As is shown in figure 7.3, they indicate that biomass on Earth may have peaked *before* the evolution of animals, during the early evolution of multicellular algae in the sea and perhaps fresh water. These models, however, were computed by number crunchers, not biologists. Is there any biological reality to them? In reviewing a previous draft of this book, Lee Kump of Penn State questioned the results of these models (unpublished). He noted:

> In these models there are three questionable assumptions: (1) the residence time of carbon in all biomass pools (prokaryote, eukaryote, and complex multicellular) is the same, (2) it neglects any sort of interaction among these biomass pools (between microbial produced and eukaryote produced) (eukaryotes provided additional food for prokaryotes); and (3) it presumes that carbon (as CO_2 in the atmosphere) limits productivity, *when in fact, even today, ultimately nutrient and water supply are probably limiting global productivity, not carbon.* Each year trees draw down CO_2, and then they release it in the fall and winter. The reason they draw it down only a few ppm is probably that growth is limited by space, nutrients, and water, not that growth is limited by CO_2. And this is the greatest level of CO_2 starvation plants have ever witnessed.

Kump goes on to say (unpublished):

> If anything, I'd conclude that biomass in more recent times vastly *exceeds* that earlier in Earth history. Today, at least half of the biomass of the planet is eukaryote, mostly trees. I also suspect that the prokaryote biomass, perhaps as large as the eukaryote biomass, is so large because of the high rates of productivity sustained by the combined prokaryote/eukaryote world. In other words, the evolution of eukaryotes added on to global biomass and perhaps, by providing an ample food source for prokaryotic heterotrophs, more than doubled the global biomass.

These are critical points. The most important knows if the addition of multicellular organisms to the planet increased biomass, and on the face of it, how could it not have? Simply looking at the biomass of life in soil and forest leaf litter gives a good indication that this must be so, for prior to common land plants this reservoir did not exist. The presence of a characteristic kind of "flat pebble" conglomerate as late as Cambrian times, and often before, suggests that land surfaces prior to the invasion by land plants in the early Paleozoic were almost soil free, without roots to help make and then stabilize soil.

NEWEST GENERATION

One of the main critiques of Lee Kump is that the combined microbial/multicellular organism world ought to have a higher biomass than did the prokaryotic world before complexity arose, and indeed the models from 2000 and 2002 do not seem to show this. However, a new study by the Franck group, published in 2006, arrived at an estimate that seemingly is in accordance with the Kump critique— but with a twist. These new results are shown in figure 7.4, and I have added on the temperature graph as it gives evidence of why some of the trends are as they are.

These new results suggest that biomass did jump with the addition of evolutionary breakthroughs – and also illustrate the "near death"—and thus highly Medean—effect of the rise of oxygen, a global poison before it became the basis of eukaryotic metabolism, around 3 billion years ago. But the most striking aspect of this graph is how quickly biomass plummets after eukaryotes are in place. Yes, with the Cambrian Explosion there was indeed a burst of higher biomass, but there were rapid drops in temperature soon after, as plants increased weathering rates to the point that silicate weathering reduced CO_2 levels drastically—and brought about both mass extinction and biomass reduction. This is a Medean effect.

Latest Pre-Cambrian Biomass—Different World Ecosystem?

The biomass peak occurs with the rise of animals. But at the same time, figure 7.4 suggests that there were more microbes as well. How could this be? Earth history may offer a clue that this is so. It comes in roundabout fashion.

Surely the most curious fossils of the Cambrian Explosion are the odd fossils known as Ediacaran fossils, named after their first recovery locality, in Australia. There has long been a debate about their

BIOMASS THROUGH TIME AS A TEST

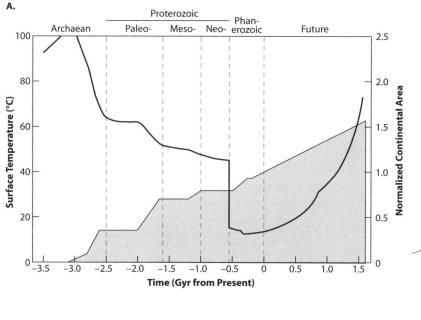

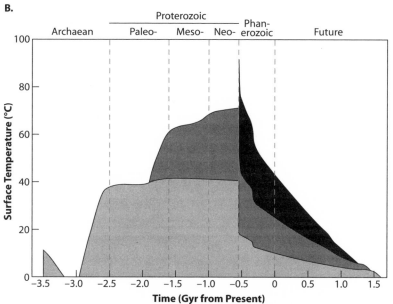

Figure 7.4. Estimates of global temperature and biomass through time. Source: Franck et al. (2006).

111

biological affinities: were they early animals belonging to still extant phyla (the original interpretation), or animals of now extinct phyla, or not even animals at all, but perhaps fungi of some sort? Sentiment has moved back to the original interpretation, that they are early animals, probably Cnidarians of some sort (modern-day corals, anemones, and jellyfish). But perhaps the most curious aspect of them is not their affinity but how they are preserved: they are found in sandstones. Such fine preservation is normally only found in fine grain shales, not sandstones; shifting sand will not for long hold a fossil imprint. So how did the Ediacaran fossils come to be preserved?

In an early class taught at the Friday Harbor labs, I assigned my students the task of re-creating Ediacaran fossils. They duly went out and collected jellyfish and sea pens (another cnidarian, looking very much like one of the Ediacaran fossils) and put these bodies atop sediment filling a container of some sort and then piling more sand on top. The inevitable rot and stench nearly got us thrown out. After two weeks we waded through the muck and looked for fossil imprints in the sand. None was seen. However, when the experiment was tried where a piece of fine nylon mesh was placed atop the sandy sediment, with the organisms carefully placed atop the mesh, and finally with sediment poured atop the whole cnidarian sediment sandwich, quite nice fossils emerged.

There is no evidence that there was a nylon coating on the late Precambrian ocean bottoms. But there may have been something like it—slicks of microbial mats. Such mats are rare today, as they are easily eaten out of existence. But in the time just before grazing animals, perhaps every sea bottom was coated with microbes, and maybe large land areas near standing water as well. If the sea bottoms were coated with microbes, with further microbial communities underneath them in the subsurface, the case could be made that biomass was higher than it is today in our animal- and higher plant-dominated world. But Lee Kump does not think so, and neither do I. However, I do believe that biomass would have been higher when the world was warmer, simply because we see such a high biomass in the tropical oceans and land areas. In such a case, it is easy to envision a world of higher biomass in the early Paleozoic, and perhaps right up until the major cooling trend of the Cenozoic, which caused global temperatures to tumble in the Miocene. I could see biomass at maximum levels following the evolution of forests, and following the evolution of wood decomposers, so that enormous quantities of the nutrients P and N made their way into the sea each season to allow phytoplankton blooms. If this is correct, it would cause there to be a peak in biomass perhaps in the Devonian. But as we have seen, devastating mass extinctions may have served to keep biomass lower than it otherwise would have been.

The planet was in a warm greenhouse from Cretaceous until about the end of the Eocene, 60 to about 50 million years ago. The Eocene was the last time that there were global forests, with palm trees existing even in high, ice-free latitudes. The long temperature slide down, caused by CO_2 reduction, was certainly produced by the emergence of continents and the withdrawal of the Cretaceous to Eocene epicontinental oceans; with more emergent land and still warm temperatures, the rate of chemical weathering caused ever more CO_2 to be removed from the atmosphere and to be locked into carbonate rocks. The rapidity of carbonate production is a byproduct of the highly evolved and highly efficient carbonate skeleton producing organisms, most importantly reef formers and calcareous microplankton of the oceans, such as coccolithophorids. It is this aspect of the Paleocene temperature drops that can be considered Medean.

With both temperature and CO_2 reduction, the Franck models project a drop in biomass, and this can certainly be recognized, even in a qualitative fashion.

IS THE EARTH'S BIOSPHERE "DYING?"

To this point we have examined past biomass and have found it to be decreasing over time. This startling new finding, pioneered by the excellence of the Franck group at Potstam, has revolutionized our view of the Earth—not as a place of increasing diversity and biomass, but as a planet slipping into old age, with lower biotic diversity and biomass. What of the future? That is the subject of the next chapter.

113

8

PREDICTED FUTURE TRENDS
OF BIOMASS

Becoming an ancestor is difficult.
—Matt Ridley, *Genome*, 1999

As we have seen, biomass seems to be highly dependent on the amount of carbon dioxide in the atmosphere and global temperature. Many things affect the amount of carbon dioxide in the atmosphere, but since the evolution of plants, biotic weathering has become one of the most important.

As the Sun continues to warm through time, it will cause a global warming that translates into increased weathering rates. The faster the silicate rocks in the crust weather, the more CO_2 will be removed from the atmosphere through the various chemical reactions that cause carbonate rocks to form. This continual removal of CO_2 will offset the solar-induced temperature increase. But there will indeed come a day when there is insufficient CO_2 in the atmosphere to allow photosynthesis. And when that calamitous day occurs, a very pronounced end to the world as we know it will begin to take place. The changes accruing will be dramatic and catastrophic to the biosphere. Let us look at new modeling that predicts the rate of biomass loss.

FORWARD-LOOKING MODELS

All of the pioneering models examining carbon dioxide and planetary temperature looked backward in time. In the early 1980s, however (with the then new understanding of the various feedback systems affecting atmospheric CO_2), a new idea suddenly dawned: not only could the various new models be used to derive estimates of

So NOT Modern?

Even w/o plants CO_2↓

OR life will adapt?

past CO_2, climate, and temperature, but they could also look *forward* in time. This is depicted in both graphs by the Franck group shown in the previous chapter. By combining the levels of CO_2 concentration, rates of weathering, and rate of continental growth through time with the known change in the energy budget of the Sun over time, highly predictable outcomes for future temperature and global "productivity"—a measure of how much life is present on the Earth at any given time—became possible using mathematical models and high-speed computers. One of the early results was the finding that the life of the biosphere (the time that the Earth could support life in any form) was finite and its longevity roughly predictable. While most scientists, if thinking about issues concerning the end of the world at all, assumed that the end would come from some cosmic crisping, even the earliest of these models showed that something far more prosaic might spell doom for the biota—dropping levels of atmospheric carbon dioxide.

The second shock was how soon this end would come: while it was known that the rising energy levels of the Sun would cause important increases in planetary temperature between 1 and 2 billion years from now, it was not known that plant life is so soon imperiled (if a half to a billion years in the future can be called "soon").

The models being used all required fast-processor computers. The model itself is a "stylized geosphere-biosphere model" that consists of values and descriptors of the four systems we have emphasized throughout this book: the solid Earth, hydrosphere, atmosphere, and biosphere. The models combine the increasing solar luminosity, the silicate rock weathering rate (discussed in chapter 5), the size of continental land surface, and the "global energy balance" (which includes the rate of heat loss to space) to estimate the amount of CO_2 in the soils and atmosphere, the mean global surface temperature any time, and the biological activity of the world at any time past or future.

As these various interactions came to be understood, it became clear as well that they could provide predictive models for the future. In the vanguard of this effort was the primary author of the Gaia hypothesis, James Lovelock. In a pioneering paper published in *Nature*, Lovelock and coauthor M. Whitfield raised the question of

115

how much longer into the future living organisms might survive on the Earth. They presciently pointed out that while too much CO_2 is a very bad thing (because of the resultant rise in greenhouse effect, and thus global temperature), too little CO_2 would be equally disastrous, for it is CO_2 that is necessary for plant growth—and without plants, the amount of life on planet Earth would be scant indeed. From this paper emerged a vibrant new area of research.

At the time of the Lovelock and Whitfield article, the carbonate-silicate feedback system had only just been proposed by Walker and his colleagues, in 1981, and it was still poorly known and little accepted. Nevertheless it was the clear to Lovelock and Whitfield that in the future, as the Sun became brighter and the increased solar luminosity gradually warmed the Earth, silicate rocks should weather more readily, causing atmospheric CO_2 to decrease. The genius of their work was in comprehending that there would come a time in the future when CO_2 levels would fall below the concentration required for photosynthesis by plants, which for most plants is about 150 ppm (in contrast, present-day CO_2 levels are about 380 ppm, but rising rapidly due to human causes). Using models similar to those employed by Berner and his group, Lovelock and Whitfield estimated that the end of plant life as we know it would occur in about 100 million years. While seemingly a staggering number, 100 million years is, in reality, a very short time for a planet that has had life for at least 3.5 billion years, and multicellular plants such as the algaelike form *Grypania* for more than 2 billion years. This result came as quite a shock.

With the publication of the groundbreaking Lovelock and Whitfield paper, the idea that sophisticated models could be used to model future events on Earth was taken up by a succession of preeminent scientists. One such group, Ken Caldeira and James Kasting of Penn State University, increased the sophistication of assumptions and model inputs. In their 1992 article titled "The Life Span of the Biosphere Revisited," published in *Nature*, Caldeira and Kasting improved the models of Lovelock and Whitfield through new terms and better values for the various parameters studied. They noted (p. 721):

[handwritten margin notes: Lots to worry about before then! Let alone revision to 1 billion yrs (off by 10x; w/ just 2 models in ~ a decade)]

The problem of the life span of the biosphere has implications not only for the future of our planet, but also for the probability of finding biologically active planets in our galactic neighborhood.... As solar luminosity increases, silicate rock should weather more easily, thereby drawing down CO_2 from the Earth's atmosphere. This feedback mechanism should tend to buffer the Earth's temperature near its present value, both in the future and in the past. Eventually, the CO_2 concentration may become so low that the megaflora existing at present will not be able to engage profitably in photosynthesis, effectively cutting off the carbon supply for the biosphere. Vanishingly low CO_2 levels would preclude further CO_2 modulated thermal buffering. The Earth would then warm more rapidly, and much of the remaining biota might be pressed against thermal barriers to their survival. Ultimately, as the sun continues to grow brighter, the Earth's surface water will be lost through photodissection and escape of hydrogen into space. The loss of surface water would bring an indisputable end to the lifetime of the biosphere.

In addition to refining many of the terms used in the original Lovelock and Whitfield analysis, Caldeira and Kasting pointed out a critical omission: Lovelock and Whitfield had assumed that plant life requires a minimum of 150 ppm of atmospheric CO_2, and this is true for the vast majority of plant species on Earth today. But Caldeira and Kasting noted that there is a second large group of plants, including many of the grassy species so common in the mid-latitudes of the planet, that use a quite different form of photosynthesis and can exist at lower CO_2 concentrations—sometimes as low as 10 ppm. These plants would last far longer than their more CO_2 addicted cousins and would considerably extend the life of the biosphere even in a world where CO_2 levels had fallen far below present-day values.

With the new calculations and values included, Caldeira and Kasting concluded their article with various estimates. Their calculations suggested that the critical 150 ppm value of CO_2 would occur not 100 million years from now, as predicted by Lovelock and Whitfield, but as much as 500 million years into the future and that perhaps some plants, using far lower levels of CO_2, might exist for as long as another billion years after that—all in all a rosier picture, or at least a world where roses could exist for another 500 million years. But Caldeira and Kasting asked the question not just in terms of when various plants would die. They also attempted to model the *amount* of life that will be present on Earth, at least as portrayed by a value known as biological productivity, the rate at which inorganic carbon is transformed into biological carbon through the formation of living cells and proteins. Here, their results were rather astounding: from the present time onward, their calculations suggested that productivity will plummet. Even though life will continue to exist, it will do so in ever-smaller amounts on the planet—not a billion years from now, not 100 million—but from our time onward. We will return to the implications of this in the next chapter.

The models used to predict the end of the biosphere continued to be improved, and even better estimates—based on newly recognized rates of weathering and CO_2 flux—continued to be published. In 1999 Franck and two colleagues improved on the Caldeira and Kasting model and looked backward as well as forward. Those results suggest that photosynthesis will end between 500 million and 800 million years in the future, and that about a billion years from now the temperature of the Earth will rapidly rise above that of boiling water.

This paper was by no means the last word. Other articles with slight refinements have appeared since, but there seems to be a convergence on a time, somewhere between 0.5 billion and 1.5 billion years from now, when land life as we know it will end on Earth, due to a combination of CO_2 starvation and increasing heat. It is that decisive end that biologists and planetary geologists have targeted for attention. But all of their graphs reveal an equally disturbing finding: that global productivity will plummet from our time onward and indeed has been doing so for the last 300 million years, perhaps. All of this will lead to a very different world, a world where life is constantly reducing its biomass.

The 2006 Franck et al. graph (fig. 7.2) was a surprise for two reasons: first, it suggests that productivity of the planet is now rapidly

118

dropping and has been doing so for half a billion years. If correct (and all analyses suggest that it is), it suggests that the age of life on this planet has been in its old age for some time. Second, the values of productivity hit 0 before another billion years is over. That gives a sense of the time left for the biosphere.

CARBON DIOXIDE AND THE END OF PLANTS

Forward-looking models suggest that the time of plants on earth is limited. How can such a bold prediction be made—that plants will disappear from our planet? The answers come from botanists studying how different kinds of plants perform photosynthesis. To understand this particular end of the world—the end of plant life—we must first examine how that now familiar character in this book, atmospheric carbon dioxide, affects the photosynthetic pathway.

Plants most markedly differ from animals in how they acquire the carbon atoms necessary for organic structure of cells and protoplasm. Whereas animals must acquire carbon from previously synthesized organic molecules (by ingesting plant or animal flesh), plants use the carbon found in CO_2 molecules and place this carbon into living material. To fuel this transformation, plants use sunlight and the well-known process of photosynthesis.

There are several biochemical pathways producing photosynthesis. The earliest photosynthetic reactions, which evolved in bacteria living more than 3 billion years ago, were surely not as efficient as later evolved systems. There has even been profound evolutionary change within higher plant lineages very late in Earth history. One of the most profound occurred only 8–10 million years ago, with the evolution of grasses and a new type of plant—a C_4 plant (as opposed to the more common types known as C_3 plants)—that can live at lower levels of CO_2 than ancestral plants can.

The formation of new photosynthetic pathways is a sure sign that the long-term reduction in atmospheric CO_2 is having a profound affect on the biosphere. The continued long-term reduction in CO_2 over the next hundreds of millions of years should produce a decisive change in global floras. Whereas the majority of vascular plants now on Earth are C_3 dicots, there should be an increasing transition toward C_4 monocot vegetation. How will this affect the appearance of global floras? One change might be the disappearance of the vast pine and fir forests of the higher latitudes, and the disappearance as well of the current mid latitude broadleaf forests and tropical rain forests composed of hard woods. Currently the majority of C_4 plants are tropical to mid-latitude grasses, and one possibility might be the changeover from a world in which great areas are largely covered by trees to an world entirely covered by vast grasslands. This possibility seems remote, however. Already there are enormously successful grasses, such as species that have evolved treelike shapes (palm trees and the various and fast-growing bamboos are examples). Yet while that tall-tree morphology may be evolved by plants using low CO_2 physiological mechanisms, it may very well be that forests, certainly as we know them now, will continue to disappear, and not just by the well-known clear-cutting and forestry of humans.

The lowering temperatures certainly reduced the rate of biological productivity. But so too did the replacement of forests by grasslands. While productive, grasslands are far less so than the forests they re-place. There would have been a net planetary reduction in biomass from this aspect alone. Add to this the replacement of the high lati-tude forests with ice caps and the formation of vast deserts of Asia and Africa, and there is a significant drop. This is strongly supportive of the Medea hypothesis.

We cannot predict the exact identity of the species present at that far distant date. But we suspect that, were we indeed able to travel freely to this far future Earth, much more would be familiar among the planet's plant life than not. There are only so many ways to make a leaf and array it so that it captures sunlight. Trees, bushes, shrubs, and grass are highly efficient at engaging in photosynthesis, honed by millions of years of evolution. We can predict that there will still be forests and grasslands. And while many (perhaps all) of the indi-vidual species will be different, the overall shapes of the animals and plants may look quite familiar, while the ecosystems themselves may function in ways quite similar to those in analogous environments today: rain forests will still be rain forest, and grassland still grass-land. But our models tell us that there will be less life on the planet by this time, and that about 500 million years from now the first great wounding of a major geophysiological system will take place. As early as 500 million years from now, or perhaps as late as 1 billion years or so into the future, the level of carbon dioxide in the atmosphere will have reached a point such that familiar plant life will no longer be able to exist.

The changeover, at first, will be in no way dramatic. All over the planet plants will slowly die. But the planet will not immediately become brown. For as one suite of plants dies, their places will be taken immediately by another cohort of plant life that may look nearly identical to those dying. Deep inside the tissues of these two groups of plants, however, fundamental processes of photosynthesis will be radically different. After this changeover, life on Earth will continue in ways probably not too dissimilar from that which came before. For a time, anyway.

There is also the possibility that plants will continue to evolve other photosynthetic pathways to compensate for lowering CO_2. In this case we can envision some sort of plant life surviving at minimal CO_2 levels. Eventually, however, even these last holdouts will die out. All models suggest that this gas will continue to drop in volume, ultimately arriving at the critical level of 10 ppm.

The time that this is projected to happen is controversial. Early models projected that this lethal blow to life on Earth—the loss of plant life—would take place in as little as 100 million years from now. More sophisticated models pushed that date back to later in the future—perhaps more that 500 million years from now—while one group suggested that, due to the biotic enhancement of weathering, there will be sufficient CO_2 for plants until about a billion years from now. Some other groups even suggest that CO_2 will hover at the critical level and never dip below it, thus allowing some minimal amount of vegetation to continue to exist on the planet. Yet even this best-case scenario produces a world vastly different from our own, and one in which there is little advanced life on the planet. Whatever the timeframe, the loss of plants will be dramatic and world changing.

It seems ironic—plants will begin to die for no apparent reason. The world will not be a hothouse. (Although it will certainly be hot-

[handwritten margin note:] Or evolve capacity to extract C from soil ∴ Solve the problem! OR bacteria decomp ↑ to release CO_2

ter than now, it may be no hotter than during the Cretaceous, some 100 million years ago.) All other aspects of the planet will be seemingly normal. Yet the plants will indeed begin to die.

The first to go will be those plants with the C_3 pathway. If plants remained in the present configurations for C_3 and C_4 species, the world would undergo a radical episode of deforestation, leaving behind mainly grasslands and species adapted for high heat and low moisture—the cactus and succulent floras and their ilk. The drop in carbon dioxide will have been occurring for hundreds of millions of years, and we can expect that evolutionary adaptations to this new environmental reality will have spurred evolutionary processes to evolve whole new types of plants in response to the lowered CO_2. But perhaps not. It may be that there will still be C_3 plants right up until they are no longer viable, resulting in a first wave that will remove the forests from the planet. By the time the first wave of hypocapnia begins to kill off plants, there may already be a global flora using the C_4 pathway.

It will not just be the land-based plant flora that is traumatized by the lowered CO_2. Larger marine plants and perhaps plankton as well will be similarly affected. Marine communities thus will be strongly affected, since the base of most such communities is the phytoplankton, a single-celled plant floating in the seas. A reduction in CO_2 will directly affect these as well as land plants. Yet the disappearance of land plants will also cause a drastic reduction in the biomass of marine plankton, even without accounting for CO_2 effects on plant volumes in the seas. Marine phytoplankton is severely nutrient-limited in most ocean settings. The influx of nitrates and phosphates into the oceans each season causes phytoplankton to bloom. But the source of this phosphate and nitrate is rotting terrestrial vegetation, brought into the oceans through river runoff from the land. As land plants diminish in volume, so too will the volume of nutrients be diminished. The seas will be starved for nutrients, and the volume of plankton will catastrophically decline. This decline will never be reversed, for even if land plants rebound at low levels, as outlined above, they will never again reach the enormous mass of material that is present in a world (such as our own) where CO_2 starvation does not exist.

On land and sea the base of the food chains as they are constructed today will disappear. The effect of this changeover from a planet coated with a veneer of plants to one without will be dramatic. Our world will no longer be recognizable to those of us living in this time of plants. The changes brought about by the loss of plants will affect and alter all four of the Earth systems: obviously the biosphere, but also the hydrosphere, atmosphere, and even the solid Earth systems. One small link in the various systems of the Earth will be damaged, and as a result all of the systems will be shaken and in some cases— the biosphere—will ultimately be destroyed by this perturbation.

The loss of plants will suddenly cause global productivity—a measure of the amount of life on the planet—to plummet. But how much? As catastrophic as the loss of multicellular plants will be, there will still be life, and lots of it. For while terrestrial plants will die off, organisms capable of photosynthesis will not. There will still be great masses of bacteria, such as the cyanobacteria, or blue green algae, that will continue to thrive, for these hardy single-celled organisms can live at CO_2 levels that are below those necessary to keep multicellular plants alive.

How much of the world's productivity is tied up in green plants? While a glance at most habitats on Earth, with their abundance of green plants ranging from grass and moss to giant trees, would suggest that most productivity would end, a more balanced view is that there would still be a great deal of productively taking place because of bacteria.

Multicellular green plants on land make up the majority of land productivity, whereas single-celled green algae in the sea provide the majority of the oceans' productivity. But there are photosynthetic bacteria in both places, as well as an unknown, but probably gigantic, biomass of bacteria in soil and even solid rock that also fix carbon. Estimates of productivity from bacterial and archean microbes alone might account for half of all the productivity of the planet.

Cutting world productivity so drastically would affect all other life on the planet, from bacteria to animals, and undoubtedly life on Earth will become far rarer. No longer will falling leaves create giant volumes of reduced carbon that makes its way into soil, the sea, and

Maybe it'll be like appearance of O_2; decline followed by a new life (imagine predictions of anaerobic scientists!)

the sedimentary rock record. No longer will coal and oil begin its process of formation. The carbon, nitrogen, and phosphorus cycles will be radically changed. There will no longer be spring plankton blooms. As land plants disappear the soil will erode, living beyond bare rock. This will, in turn, perturb the hydrological cycle, and even the pathway of liberalization on the planet. Giant transfers of carbon between the various land, ocean, and sedimentary record reservoirs will occur.

The disappearance of plants will drastically affect landforms and the nature of the planet's surface. As roots disappear and surface layers become less stable, the very nature of rivers will change. The large, meandering rivers of the modern era date back, at most, to the Silurian period of some 400 million years ago, when land plants first colonized the surface of the planet, for it takes root stability to maintain the banks of meandering rivers. When plants die out, or are not present due to slope, soil, or other inopportune environmental conditions, a different kind of river exists—braided rivers or streams, the kinds of flows found on desert alluvial fans or in front of glaciers, two types of environments not conducive to rooted plant life. This was the nature of rivers before the advent of land plants and will again be the way that rivers flow when CO_2 drops to the plant dieoff threshold.

The loss of soils will be no less dramatic. As soils are blown away, they will leave behind bare rock surfaces. As this condition begins to occur over the surface of the planet, it will change the albedo—the reflectivity of the Earth. Far more light will reflect back into space, thereby affecting the Earth's temperature balance. The atmosphere and its heat transfer and precipitation patterns will be radically changed. Blowing wind will begin to carry the grains of sand created by the action of heat, cold, and running water on the bare rock surfaces. While chemical weathering will lesson due to the loss of soil, this mechanical weathering will build up an enormous volume of blowing sand. The surface of the planet will become a giant series of dune fields.

While this event could signal the final extinction of all plant life on land (and perhaps in the sea as well), it is more likely that a long period of time (perhaps in the hundreds of millions of years) will ensue in which CO_2 levels hover at the level causing plant death. As the levels drop to lethal limits plants die off, reducing weathering and allowing CO_2 again to accumulate in the atmosphere, once again allowing any small surviving seeds or rootstocks to germinate and, at least for some millennia, to flourish at least at low population numbers. As plant life again spreads across land surfaces, weathering rates will again increase, CO_2 again will be reduced, and plants again will die off.

GREEN PLANTS AND OXYGEN

Animal life is dependent on an oxygen atmosphere. There are no animals capable of living in zero- or even low-oxygen conditions. With the loss of plants, what happens to atmospheric oxygen? While some scientists have thought that the loss of plants will have little effect on atmospheric oxygen values, new studies suggest just the opposite. The loss of plants will shut off the major oxygen-producing pathway on the planet—photosynthesis. But the loss of plants will have no affect on the most important oxygen "sink"—the oxidation of dead matter on the surface, and volcanic gases emanating from the interior of the Earth. It is the latter that will most rapidly deplete the oxygen supply. A recent calculation by astrobiologist David Catling suggests that by about 15 million years after the death of plants, less than 1 percent of the atmosphere will be oxygen in contrast to the 21 percent volume that the world has today.

THE CULPRIT

From the last two chapters it appears that life is to blame for both short-term and long-term reductions in biomass, and life will be to blame for its own death. If an enlarging Sun exterminated life, that argument could not be made—but this is probably not the case. It is life—through the formation of massive amounts of calcareous (limestone) skeletons now perched on land (and out of the carbon cycle), as well as through the biotic enhancement of weathering—that has caused the long-term decline in CO_2. Life is to blame. No, Medea is to blame.

Seems like accelerated weathering but would've happened anyway.

9

SUMMATION

Let us sum up—in the shortest chapter of all. Three hypotheses have been presented. The first, the Gaia hypothesis (Optimizing), promotes the idea that life makes conditions better for itself. The second Gaia hypothesis (Self-regulating or Homeostatic) posits that life maintains conditions that, if not optimal, certainly stay within habitable bounds. Third, the Medea hypothesis suggests quite the opposite—that life, and future life, limits itself in any number of ways, and does so in no small way by causing positive feedbacks in various Earth systems necessary for life. A number of specific tests were proposed early on. They included the following.

1. *Does the history of diversity support the Gaia hypothesis?* It should show ever-increasing diversity through time. It does not. Diversity of animals and higher plants seems to have been in a steady state for more than 300 million years since the evolutionary conquests of land, with this long-term value depleted on occasion by mass extinctions. Second, we do not know what the diversity of microbial life was prior to animals, but it was likely higher. The almost complete loss of stromatolites with the Cambrian Explosion indicates that microbial biomass was certainly higher prior to animals, and it may be that biodiversity was as well.

2. *Does the history of biomass through time support the Gaia hypothesis?* It does not. Model results indicate that biomass on Earth peaked some 1 billion to perhaps 300 million years ago and has been diminishing since. Since two main factors affect biomass values—temperature and atmospheric carbon values—we should look to these two. Temperature has remained fairly constant, but carbon

[handwritten margin note, left of first test:] · Not clear; stasis & recovery = Gaia or Medea?

[handwritten margin note, left of second test:] Weak — could be cyclic on long term.

values have plummeted as CO_2 has been removed from the atmosphere by increased carbonate silicate weathering by plants, as well as the increased efficiency of carbonate skeleton production by animals and plants, microbial to macro in size. Both of these factors causing reduction of CO_2 were caused by life. This is not in accordance with predictions from either Gaia Hypothesis

3. *Will the future of biomass show a steady decline through time up until the loss of oceans?* The Gaia hypotheses predict that life will extend the life of the biosphere. But model results suggest quite the opposite—that, through the removal of CO_2, life itself will cause a shortening of the timeframe within which the Earth can sustain surface life. While microbial life might still survive following the loss of plants, an oxygen atmosphere, and finally the oceans, there is no consensus that the Deep Microbial Biosphere can withstand the loss of all surface life.

compared to ?

speculation on billion yr extrap w/o novelty

4. *Do individual events during the life of the biosphere show evidence of Gaian influences?* Since the main life-related events on the planet after life's first appearance include the oxygen rise, the Snowball Earth events, the Cambrian Explosion (appearance of animals), and the various Phanerozoic mass extinctions, this question can be proposed in the light of these events. Each, however, as we have seen above, produced a reduction of biomass at the time.

& then recovery

In summary, the four points above seem to me, at least, sufficient to falsify the Gaia hypotheses. Does this mean that the Medea hypothesis is correct? Not necessarily—the hoary "more research is needed" is all too true. But the evidence at hand certainly points to it being a better descriptor of how life works than the Gaia hypotheses.

No - shows sometimes M or G (scale dependent)

This could be the end of this book. But it was never my goal just to lay siege to Gaia, replacing that benevolent mother figure with a silent murderess. Let us move to two final chapters: the first dealing with environmental implications, and the second, a brief essay on what we might do to save our species from extinction.

10

ENVIRONMENTAL IMPLICATIONS AND COURSES OF ACTION

What a trifling difference shall decide which shall
survive, and which shall perish.

—Charles Darwin

This was the hardest chapter to write. It is far from my comfort
level (science), for it called for philosophy and meditation on the future, and I therefore beg the reader to forgive my undoubted inele-
gance here, for philosophy and meditation are in short supply in my
makeup. I will try a short summary of the main point: the implication of viewing life as Medean rather than Gaian requires a paradigm
shift in our worldview. *We must change from being witlessly destructive
life forms to being consciously active anti–Medean life forms.*

This is a big change (one might almost call it an evolutionary shift,
but that is not quite right, either) that takes us out of philosophy/
hand-wringing and into action in ways and for reasons that go far
beyond even the level of directives emanating from the Al Gores and
others so concerned about global warming, global pollution, global
poverty. We are in a unique position—in comparison to the history of
biology on the planet—and our survival (if we are indeed fit enough)
depends on embracing this paradigm shift and taking action. It is
hard not to fall into hyperbole here, but the stakes are real enough.
We must not become part of nature. We must overcome nature.

In these final chapters, I will try to integrate the overriding hypothesis of this book—that "nature" is inherently "Medean"—with
two other ramifications: in this chapter, I will discuss the effect of
my central hypothesis on what is called environmentalism, and in

Is human consciousness the Gaian response to Medean CO_2 loss?

↙

MANY threats more sure & imminent than CO_2 loss in 100s millions of yrs!

the subsequent (and final) chapter, I will propose some necessary, planetary-scale "adjustments" that will be needed if we are to extend the lifespan of the biosphere. But first we must look at environmentalism, and the implications of how it might change if the Medea hypothesis is indeed correct.

HUMANS AND THE NEW FREEDOM OF UNLIMITED TRAVEL

It is only in the past three to four generations that a significant portion of humanity has routinely traveled the globe with such freedom and frequency. As India and China so rapidly move into a vast middle-class society, the number of globe-trotters is set to double, then treble and more. Boeing and Airbus are licking their chops, and those trying to slow global warming are close to acknowledging defeat, since there is no known way to meaningfully reduce the emissions of a kerosene-fueled jet engine, and jet travel is and will remain one of the greatest polluting activities on Earth, as we humans jet forth in ever greater numbers.

[handwritten margin note: huh ?! Hardly 1st prob.]

While many of the air-traveling humans are and will be going to view the great sights of human civilization, many more are and will look for the exact opposite: places where there is no civilization—the wild places where Maurice Sendak's wonderful beasts might still live in the pristine grandeur of prehuman contact. Unfortunately, most people who search for the untouched are vastly disappointed. While there are indeed seemingly unspoiled bits of ancient Earth still to be found, they can only be reached by going through ever-larger human cities, the great jet hubs, or other tracked wastes of civilization. A realization that so many of us have come to upon reaching or transiting these cities and spreading human habitations is how polluted and tagged by garbage they are, particularly in areas once referred to as "third world countries."

Is there any place on Earth not yet desecrated by the plastic grocery sack, that ubiquitous signpost of human progress? Is there any air in the world worse than that in Mexico City, or Bangkok, or Moscow, among so many others? It takes real money to enact a Clean Air Act, Clean Water Act, or their equivalents, and most countries on Earth are not yet rich enough to afford them. The developed world wrings its hands about lowering carbon emissions, while all the world knows that the fate of the atmosphere is in the hands of China and India, and soon also Brazil, Mexico, Pakistan, and the North African Arab states, all ground zeros of human population and industry increase. If money can be spent on cars or pollution controls, which wins? Two cars in every garage is rapidly becoming a global reality, and the future shown in the *Blade Runner* ilk of movies seems ever nearer. Of course we seek a way out. But that way seems ever more difficult to find, especially when burgeoning cities such as Bangkok and Mexico City, both beyond 10 million humans and counting, have populations where over half the inhabitants are under thirty; in Cairo, over half are under twenty.

Yes - at least to respond to limits (neg feedback); form is debatable

What is our hoped for alternative? Why, back to nature, of course. We humans living in harmony with nature: all of our creature comforts, instant communication, and rapid global transportation while living amid tall trees, grasslands, and the biodiversity and abundance of wild animals and plants that existed in prehistory.

Hope for our planet's future (which really by extension is a hope for its life, rather than the rocks and water that this life lives on, in, or over) comes from an increasingly global recognition that there are environmental problems on this planet that are being exacerbated by the ever-increasing human population. Hope too comes from many areas of human endeavor and thought spurred to action by that realization—the millions of humans dedicated to ensuring that the Earth is a better place for our children than it was for us, with cleaner air and water, less pollution, and more visible wildlife and "natural" areas replacing human blight and workings.

Despair, on the other hand, is also inescapably linked to that hope, like night and day. It comes from the realization that so many of us *still* do not accept that there is a problem or, more depressingly, understand the challenges but act against the needs of others through ignorance or, more commonly, greed for money or power over fellow humans and other organisms.

The greatest hope for dealing with the environmental challenges of overpopulation comes from the kind of thinking, action, and consciousness-raising we typically find under the broad banner of environmentalism. It often seems like the only bridge to human and planetary survival. The challenges to building such a bridge are mainly due to the realities of feeding, clothing, warming or cooling, employing, and transporting the ever enlarging population of humans.

What is the philosophical foundation of environmentalism, and how does that core belief translate into action? Just how could embracing this philosophy ease or even ensure the future of humanity on this rapidly shrinking planet?

The main message of the environmental movement is that if we "return to nature," or turn the world back to its state before humanity evolved—in other words, stop pirating the Earth's natural processes and resources for our short-term benefit and instead try to return to something resembling our relationship to the planet before we "took control" of nature—the Earth will eventually clean up our mess and save us from ourselves. We need only take a bridge back to how things were before widespread civilization bloomed on the Earth—and of course try to keep the best of both worlds: keep civilization, but also maintain a positive relationship with the ecosystems of other species and with our environment/climate in general.

As currently practiced, environmentalism is a large movement inclusive of many laudable goals: conservation—of fuel, species, resources, habitat; activism—political change through voting green; management—of the large areas "saved" from development; and protection—of the many species currently at such low numbers that only laws will enable their survival.

But what is the overall goal of environmentalism? Taken all together, it is an admission that human civilization is creating a wreckage of the nature that existed in prehuman times, or at least in preindustrial times. It is the dream of a return to the pristine past, but also more than that.

The environmental movement is increasingly proactive. Its most extreme form, radical environmentalism, advocates eco-terrorism. Mainstream environmentalism attempts to return things to the state they were before humans entered the picture, for one main philosophical assumption is that humans are not part of "nature," at least

131

as far as I can tell by looking at environmental writings available to me during the writing of this book. For instance, the renowned environmentalist Barry Commoner stated in his "third law" that any major man-made change to a natural system is likely to be detrimental to that system. Commoner is far from alone in this view that humans are somehow "separate" from nature. Thus, the impetus to return things to the prehuman state is one direction advocated by some in the environmental movement.

Unfortunately, practicing this can lead to wild contradictions and unforeseen consequences that are sometimes worse than the practices trying to be improved upon. There are so many examples. In Yellowstone Park, for instance, wild wolves have been reintroduced, yet we also allow unlimited snowmobile use in winter, and automobile use in all seasons, such that the snowy, mountainous terrain within and just outside the park becomes as polluted as Bangkok. Our efforts in Yellowstone are so contradictory: we try to return the environment to how it was two centuries ago, yet we will not allow the frequent wildfires that were and are so necessary for keeping the park in ecological equilibrium. Even placing it as an island surrounded by fenced ranchland makes it susceptible to the same reductions in biodiversity first demonstrated on oceanic islands four decades ago in the great book *The Theory of Island Biogeography*, by Robert H. MacArthur and Edward O. Wilson. Replacing grizzly bears in Montana and Alberta, in the same regions in which numerous summer cottages have replaced large ranches; ceasing the tiger shark eradication program while more than a million snorkelers take to the bays on Oahu as so much shark bait; the list goes on and on. We want it both ways, and worse, we seem to be fundamentally confused about what exactly we want our relationship to nature to be.

Africa is the same. The vast game preserves in Kenya, South Africa, Namibia, and others have become oases of ancient Africa. But the preservation of elephants, so necessary only two decades ago, has now led to elephant overpopulation that threatens the food of the human overpopulation. The elephants leave their preserves, trample crops, as is their blundering wont, and get "culled." Some balance.

These are zoos, not a return to nature, and as such, they are to be fought for and saved. But we humans cannot return any significant part of Africa back over to the vast herds so long as there are so many humans farming or grazing those same grasslands. And because of this, there will never be spaces large enough on our planet to allow the kind of geographic separation necessary for the formation of a single, new large mammal species. We have entered a time of unnatural selection: survival of the firmest and tastiest fruits and vegetables, and in the case of animals, survival of the stupidest and friendliest, until just before we eat them. There is no doubt that this is an unhappy state of affairs.

It is thus a goal of environmentalism to "fix" this situation. The scientific framework of environmentalism harbors a firm belief that the many Earth systems that existed throughout the history of this planet—the carbon, nitrogen, phosphorus, and sulfur cycles, among many others—have been perturbed by human activity to the detriment of not only other species but ourselves. Moreover, the framework of environmentalism makes an equally firm prediction that if we could somehow put these cycles back to work in their pristine state and return the biota to its wild state, our species would transcend, and ultimately survive.

HUMANS AS PART OF NATURE

We evolved from another primate species. A hominid species, some kind of *Homo erectus*, with some mucking about became *Homo sapiens sapiens* (we are even our own subspecies!) some 200,000 years ago, and there is evidence of major brain changes with intellectual ramifications only 35,000–40,000 years ago. Since then, however, we seem to have become somewhat evolutionarily stabilized—at least in morphology. We were certainly brought about by natural selection in thoroughly unoriginal ways, being one of the more recent species to have evolved on the planet following more than 3.5 years of evolutionary practice by nature.

We were part of that nature. Times were tough. Climate bounced around in rather unpredictable ways, and only with a calming of climate that occurred some 10,000 years ago, a long warm period, did

133

humanity step away from the threat of extinction due to small numbers. With the calm came agriculture, and we never looked back.

So when can it be said that we stepped out of the natural world to become exceptions to the ecology of the planet, if that particular view indeed has any truth to it?

THE PHILOSOPHICAL UNDERPINNINGS OF ENVIRONMENTALISM

The Greek word "philosophy" means literally the (filial) love of wisdom. There is a new discipline named environmental philosophy that attempts to interject wisdom into the underpinnings of environmentalism. There is a large literature exploring the foundations of modern environmentalism. In a good summary of this subject, Eric Reitan wrote the following:

> One of the most recurring themes in contemporary environmental theory is the idea that, in order to create a sustainable human society embedded in a flourishing natural environment, we need to change how we think about our relationship with nature. A simple change in public policy is not enough. Modest social changes—such as increased use of public transportation or a growing commitment to recycling—are not enough. Nor is environmental education that stresses the dangers of current practices and the prudence of caring for the earth. Even appeals to moral duty—obligations to future generations and to the fellow creatures with which we share the planet—are insufficient. What is needed is a change in our worldview. More specifically, we need to change our view of nature and of our relationship with nature.

I certainly agree with the last sentence (although in a way that, I am sure, would horrify the author). As to a "flourishing natural environment," I am curious if the author of this and so many other such texts really wants such a thing. In my long work as a field geologist and field marine biologist, I had to survive in a number of the few remaining such natural environments, one amid the nocturnal, outer coral reefs of the Indopacific during two decades of researching the

134

chambered nautilus in situ, the other in the grizzly and large black bear territories of the Canadian cordillera, Alaska, and the Queen Charlotte islands. In both cases it was a requirement to travel armed, for protection, and even so I came close enough to getting eaten a number of separate times that I began to feel empathy for those otherwise to be hated "pioneers" who went out of their way to kill all the bears of North America, and who are now so vigorously attempting to kill every large shark in the sea. The urge is natural enough—anyone with children will try to reduce ambient danger, and the fact is that humans were getting eaten on a regular basis not so long ago.

A return to natural conditions includes returning the human-eating predators back to their original numbers. Do we really want to see our children in danger of being eaten, as they were for all generations up to about ten or twenty ago? And if you remove the top predators but try to get back everything else, the result is as unnatural (in its own way) as is a human city street environment. So there is a large dose of hypocrisy here. What is really being asked for is the equivalent of a golf course: lots of trees again, and a return by some of the smaller animals that lived in the original forests (which would be removed to make way for the duffers), but just some of them. Any man-eaters, mosquitoes, and/or large herbivores that might muck up the greens would not be allowed.

One of the most influential current movements of environmentalism is deep ecology. The website Greenfuse (http://www.thegreen fuse.org) is the basis of the following discussion. Deep ecology was originally developed by a Norwegian philosopher, Arne Naess. It has grown into a worldwide movement of considerable influence. Naess presents a set of principles that he invites people to integrate into their own personal philosophy of life:

Deep Ecologists emphasize that human beings are only part of the ecology of this planet, and believe that only by understanding our unity with the whole of nature can we come to achieve full realization of our humanity. Deep Ecology believes that all organisms are equal: Human beings have no greater value than

135

any other creature, for we are just ordinary citizens in the biotic community, with no more rights than amoebae or bacteria.

This certainly sounds reasonable. But the paradigm shift described at the start of this chapter deals exactly with this point and turns it on its head: we are not ordinary citizens. We are the only hope to keep Earth life alive.

THE DEEP ECOLOGY PLATFORM

One branch & not terribly active!

The eight points of the "Deep Ecology Platform" (Naess 1989. p. 29) can be paraphrased as follows:

1. The flourishing of human and nonhuman life on Earth has intrinsic value. The value of nonhuman life forms is independent of the usefulness these may have for narrow human purposes.
2. Richness and diversity of life forms are values in themselves and contribute to the flourishing of human and nonhuman life on Earth.

VITAL

3. Humans have no right to reduce this richness and diversity except to satisfy vital needs.
4. Present human interference with the nonhuman world is excessive, and the situation is rapidly worsening.
5. The flourishing of human life and cultures is compatible with a substantial decrease of the human population. The flourishing of nonhuman life requires such a decrease.
6. Significant change of life conditions for the better requires change in policies. These affect basic economic, technological, and ideological structures.
7. The ideological change is mainly that of appreciating life quality (dwelling in situations of intrinsic value) rather than adhering to a high standard of living. There will be a profound awareness of the difference between big and great.
8. Those who subscribe to the foregoing points have an obligation directly or indirectly to participate in the attempt to implement the necessary changes.

There is little to argue with here. Certainly the key point is a need to reduce human population numbers. But while laudable, how much of this is practical? Let us return to the Reitan article and examine that point about pragmatism, for Reitan concludes his very thought-provoking piece as follows: "Human beings evolved in the natural environment that we are presently transforming. We evolved to be dependent upon that natural environment for our physical as well as psychological sustenance. Our actions amount to a destruction of much upon which we depend, and are therefore self-defeating in a very straightforward way. The worldview that impels such actions is therefore *pragmatically* false."

I have added the italics in the last sentence, for it is an important aspect of this entire discussion. Pragmatism is grounded in reality, and the most basic reality of our globe is the presence of over 6 billion humans, all large animals not only changing the environment around themselves, but leading to changes everywhere. Buying anything made in China if one resides in North America or Europe causes this effect, as does flying, eating transported food, and on and on.

Preserve as much of the still pristine natural places as we can. Who could argue with that sentiment? *Conserve.* Who besides those beholden to business interests can argue with that one? Simply by replacing the entire fleet of the world's automobiles with Toyota hybrids, replacing all light bulbs with low-energy fluorescents, and banning all air travel would rather quickly stabilize atmospheric carbon dioxide values, now rising so quickly that our planet is rapidly returning to an Eocene-like atmosphere that will cause all ice caps to melt in the next two to five millennia and therefore cause a rise in sea level of more than 240 feet. That is a seemingly pragmatic action, and yet any reasonable person recognizes that humanity will certainly not do such a prudent thing anytime soon. The Chinese and Indians, among so many others, have yet to know the joy of an eight-cylinder SUV rolling down pristine interstate highway systems connecting all of Asia, but that is certainly the future of that part of the planet.

Pragmatic environmentalism is a hard road. So too is being brutally honest about who we are as Darwinian organisms in a Medean world. We won, and every other species would "love" to be in our shoes. It is in our natures, all of us Earth life, and perhaps, all life.

SOME RULES OF PRAGMATIC ENVIRONMENTALISM

As I peruse the vast literature I am continually amazed by the vast and assorted intelligences that have contributed to this field. But I would love to punch the skeptical environmentalist, the *Fox News*–friendly Bjorn Lomborg. And there *have* been real "environmental" victories. Great pioneers such as Barry Commoner point to the golden victories of the Nixon years (!!!), when the Clean Air Act, Clean Water Act, and Endangered Species Act all were installed as the law of the land in the United States, when DDT and assorted other chemicals were banned, when awareness was raised. The end result was the conservation of many species that otherwise would have gone extinct. But the real political reason behind these seeming acts of "altruism," a trait suggested to be nonexistent in Darwinian organisms, is that every one of them materially improved the health of us humans. Even the Endangered Species Act, the most altruistic of the lot, has a very pragmatic core, for the most endangered species are precisely those that serve as proverbial canaries in the mineshaft. They are signs that human pollution and disruption in given geographic areas have reached levels requiring a change in human behavior.

So what might be some really pragmatic rules to a new environmentalism? First, I would advocate that humans should accept that technology is here to stay—that we won an evolutionary lottery from a biosphere that would (and will) kill us if it could, and that we should make the best of things, and I do mean the best. But pragmatically. The first step is a new philosophical underpinning to environmentalism. We do not want to go "back to nature." We have done that, and left to its own devices, nature has shortened and will continue to shorten the lifetime of the biosphere. So let us look at human activities that reduce biosphere productivity (a metric of Medean meddling) and change these as a first step.

I would suggest that the two most dangerous human activities that "harm" the environment are climate change and warfare, and for the latter, the wider and longer the war, the more devastating the effects. The tall columns of smoke rising from the fired oil wells and refineries in the first Gulf War are certainly memorable examples; other examples are all too well known. The worst-case scenario would be wholesale exchanges of nuclear bombs, using hundreds to thousand of warheads. Perhaps that is where we are indeed going, and that would certainly qualify as a supreme Medean event. Radiation poisoning through fission or fusion bomb attacks would the greatest human Medean effect, and the current arming of the Muslim world, whose regions show the highest rate of population growth on the planet, is the greatest threat.

The way to stop warfare is to increase the standard of living all over the world. This will only happen through even higher rates of Medean activity by humanity—burning more coal and oil, building more nuclear power plants that produce plutonium wastes, building more "infrastructure" such as continent-spanning highways where none now exists. None of this will help take us back to "nature," and in some cases it will further reduce biomass and diversity over the next few centuries. There is no stopping this, but the alternative seems to be long-term privation of some societies, which inevitably leads to warfare in a world where every state can arm itself with a nuclear arsenal. Though we should save as much of nature as we can, of course, it will probably not be much. Yet, lots of golf courses and game preserves are still better than a radioactive world.

As for climate change, we must stabilize the carbon dioxide level of the atmosphere, as every right-thinking human already knows. It cannot go too high or too low. Easy in theory; perhaps impossible in practice.

PHILOSOPHICAL CHALLENGES

The obvious and specious conclusion that could be drawn from all of this is that Medea gives permission to her children to continue their rape and pillaging of the planet. Nothing could be farther from the truth. To stop a biomass-deadening greenhouse extinction we will have to lower the trend toward higher atmospheric carbon dioxide. We must preserve as much of the green areas of the planet as possi-

ble to keep the globe oxygenated. We must reduce toxins that themselves are poisons to the biosphere. We must become anti-Medeans
in our actions—which would make us Gaians, a faintly funny out-
come of all of this, at least to me, irony being in short supply these
days.

Exactly—
we are
the Gaian
response to
current Medean

11
WHAT MUST BE DONE

In the short term (nothing about Ch 7-8; incoherent set of concerns; solutions NOTHING about Medea only about human screwups

And the rain was upon the Earth 40 days and
40 nights.
—Genesis 7

We are in a box. Ultimately it is a lethal box, a gas chamber or fryer, depending how things work out. If we as a species are to survive, we will have to do a Houdini act.

In this chapter I will suggest a series of engineering feats that will have to be accomplished.

WHAT WILL HAPPEN

My friend and past coauthor Don Brownlee, the man who successfully guided a spacecraft far out into the solar system and retrieved bits of a comet in the now famous NASA Stardust mission (itself named for a line from Joni Mitchell's song "Woodstock," which pretty much pins Brownlee's age and leanings), loves Hawaii, specifically Maui above all places (at least, from what I can gather). Yet it was Don who pointed out to me the fact that all of life on Maui, and each of the other Hawaiian Islands, is quite doomed. The history of those islands has been one of active volcanism as each passes over the active "hot spot" that currently still builds the "Big Island," Hawaii. Yet as each of the islands slowly moves on in a northwesterly direction, the lava spigot is turned off beneath it, and it begins to sink and erode ever lower and smaller. This takes millions of years, but one only has to island hop to the northwest to see the veracity of this process. Currently there are thousands, perhaps hundreds of

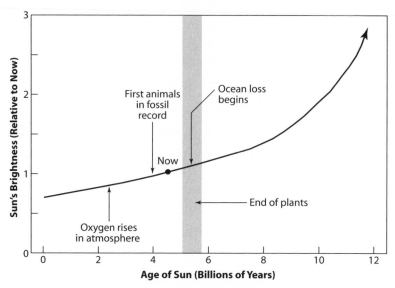

Figure 11.1. A timetable of the various "ends" of the world. Source: Ward and Brownlee (2003).

thousands, of species on the islands, yet as each island finally sinks beneath the sea, its entire endemic species, one by one, dies out. No one (beyond people like Brownlee, for whom large numbers and distant times in both the past and the future are understood with equanimity) seems too worried. But all of that fauna and flora is doomed—except for those that might have the good fortune to float off on a log to wash ashore on still extant land, provided conditions on this new land are favorable for survival.

The same is true for the Earth and all of its inhabitants, save the one species that is capable of floating off into space on our version of that floating log. To survive, we will need to move off our planet—or move our planet. But even that is but a temporary measure.

The conditions that will require us to leave are the enlarging Sun and decreasing carbon dioxide levels. But we could always get killed off long before those two aspects, still at least 500 million to a billion years into the future, take hold. The most likely ways to die—not of planetary old age but of catastrophic accident—are via asteroid impact, gamma ray burst, or nearby supernova. But far more likely will

Exactly

right

be extinction either from full-scale nuclear exchange or from yet another greenhouse extinction episode. What will we have to do to avoid these fates, and keep our species going beyond hundreds of millions of years, into the billions of years range? Our "out" is a combination of common sense, political will, and engineering on a massive scale that is possible only with a global rather than fractious civilization. Utopian pipe dreams, of course. But then extinction lasts a long time.

There is some dark irony in what must be done. In the near term we must reduce atmospheric CO_2. Then, in the long term, we must move to keep CO_2 from falling too far. But with significant engineering, both are readily possible.

THE NEAR TERM

We live in a time with high continents and few inland or "epeiric" seas. But then again we live in a time with ice caps, and such times have been few in the Earth's long past. The geological record clearly shows that for most of geological time, sea level was far higher than it is today. And because of this, science knows a great deal about the geology, geography, and biology of a more flooded Earth. I believe we are in for an inevitable return to that state, one not far in our future, but very near—perhaps as near to us as the builders of the pyramids are behind us.

There is no doubt that planet Earth is radically changing through global warming. Those resisting this conclusion are doing so for political, economic, or deficit-of-intelligence reasons, not as a result of scientific facts. There is also no doubt that to allay the unquestioned and global changes confronting this planet and all of its inhabitants over the next decades to centuries will require vast outlays of the Earth's technological, human, raw, and monetary resources. It is becoming increasingly clear that humanity is in a fight it can no longer win. On battlefields in such situations, triage is undertaken: save what you can, but put your resources behind things that can be saved. If the Earth as we know it—a planet with ice caps, extreme temperature changes from pole to equator, and major weather patterns varying with latitude and even longitude because of this latitudinal temperature gradient—cannot be saved in its current state, what is the best that we can hope for as a species? In other words, if we cannot stop global warming, how can we make the best of the situation?

There are now many books about the supposedly worst aspects of a new climate; most seize on storms and violent weather patterns. Yet surely the sea's slow rise to a point of, at first, 20 feet (with the melting of the Greenland ice cap) and then 240 feet (adding the Antarctic ice caps to the global ocean) *above* current levels within two to four millennia will quite overshadow every other effect, and indeed, with the warming will eventually come a calming of weather, not an intensification. Global warming heats higher latitudes; the tropics are already as warm as they will be. The changes to plant life will be enormous, but the changes to human society will be devastating and catastrophic unless planning is begun now.

CO_2 levels are rising by 2 parts per million yearly and that figure is accelerating. Levels of 1,000 ppm in the past usually, if not always, led to ice-free worlds, and with our current level at 380 and climbing, the dangerous 1,000 ppm level would be reached in 300 years at most, or in as little as 95 years according to University of Washington climatologist David Battisti. But many climate specialists believe that 2 ppm/year change can only accelerate: on average, China puts three coal-fired generators on line each week, and India is scheduled to become the world's most populous country by 2050, with a greater population than China and the United States combined. The emerging middle class in India is only slightly behind that of China, and as in China, burning coal is its vastly preponderant energy solution. Not far behind is Brazil, also with a large emerging middle class—the group that sees cars as a necessity for life. For those reasons alone a carbon dioxide level of 800–1,000 ppm in about a century from now is certainly possible, and perhaps probable. The last time CO_2 was so high was in the 55-million-year-old Eocene—and the climate of that world was warmed such that palm trees and crocodiles could be found far north of the Arctic circle, an Arctic, like the Antarctic of the time, without continental ice cover. We are rapidly going back to the Eocene.

It is the past that allows us to predict the future: our planet has been in an ice-free condition many times in the past, thus allowing very accurate predictions about the oncoming geography, climate, and distribution of biota. But humans have never existed in such a world.

It is within fifty years that the first really important societal effects will be experienced. This will come from two things: the first sea-level rise on ancient coastal cities, and that same sea-level rise on deltas. At this time sea level rise will still be seemingly small, on the order of 50 cm, or as much as 1 m if the Greenland ice sheet melts faster than currently (and optimistically) suspected. While coastal cities in highly industrialized countries will fight back using technology in a fashion and stance similar to that of the Dutch and Venetians today, there will be many cities where such an approach will not be possible. The greatest effects on these cities will be the loss of underground infrastructure, and the collapse of poorly reinforced buildings. This will also be the time that the last pack ice disappears from the Arctic, and the old dream of a Northwest Passage will have come true. The economic, societal, and biological consequences of an ice-free Arctic will preoccupy our species.

By A.D. 2100 the rise of sea level will have begun in earnest. There will have been a 1 m rise in sea level. All coastal cities will now be fighting the sea, but fight they nevertheless will. Yet this marks an extreme turning point. No longer can there be denial of the flood to come. This time also corresponds to a major migration and/or extinction of plants. A new, stable climate regime affects all growing areas save those already in the tropics. This is also the time when the hydrological cycle undergoes irreversible changes toward aquifer salting along coastal areas and far inland along rivers, creating local extinctions among wetland organisms accustomed to fresh water only. Food production is everywhere affected. Global population has just hit 9–11 billion.

By 2150–2300, the melting of the Greenland and West Antarctic ice sheets will be well under way and hopefully not completed: while all attention has been focused on the Greenland ice cap as the first to go, there is now good evidence that one of the larger of the several distinct ice caps in Antarctica is also threatened, and melting. Thus there is a strong possibility that early melting will be faster and will produce more liquid water than earlier conceived. Estimates vary regarding when this melting might be complete; perhaps it will be as early as 2150, or perhaps slightly more than a millennium after that. Here we see the level of the sea rising to 20 m above the present level, thus dooming all coastal cities. This period will correspond to major human migration, famine, and, undoubtedly, warfare.

Ultimate submergence will occur from 2500 to 5000. Here we arrive at the end point of sea level rise. The ultimate date depends on the rate of the melting of the Antarctic glaciers. With climate change and sea-level rise, a variety of geological processes will change. These range from the kind of clay being produced, to rates and processes of weathering and landforms, to regional climate patterns. The nature of nature will have radically changed.

The fate and use of the many coastal cities following the total sea-level rise will not be pretty. Only skyscrapers of twenty-four stories or higher will still be present in those cities exactly at sea level.

The newly drowning coastlines will radically change farming practices, crops, and species used. Wheat and other cool-weather grains will shift toward higher latitudes; current temperate areas will have to be converted to mainly tropical crops. Water-distribution patterns will be radically changed. The loss of all deltas and low-level rice-growing areas will require this most important of all Asian food staples to be completely relocated.

Of great importance will be the nature of crops from high latitudes. Today, even though very near the Arctic region, the Matanuska Valley of Alaska, near Anchorage, produces enormous amounts of vegetables even in its short, temperature-dictated growing season. Because the summer days never get dark, the plants have nearly twenty-four-hour growing intervals from available sunlight. In the new world, warmer temperatures will allow crops to start earlier and end later.

The two most important new land areas resulting from warming will be the newly de-iced Greenland and Antarctica. Even with sea level rise, these will be important and large new agricultural areas—as long as major engineering keeps them from having large inland seas. The weight of the ice sheets for so long on their surface has caused the land surface to be depressed. It will rebound with ice loss, but nowhere near as quickly as the rate of ice melting. New scientific papers show that these vast inland basins will fill with seawater. It turns out that the entry points for seawater are narrow: for both Greenland and Antarctica, large dams along these entrant points can be constructed to keep the seas out but allow the inland areas to produce Great Lakes–type inland lakes instead. These will become the two largest freshwater lakes on the planet, as long as this engineering feat is completed.

The sea-level rise, when finished, will take us back to something akin to the Late Cretaceous geography. This was a time of major epeiric, or inland, seas. A large sea will occupy the interior of North America, the Amazon Basin of South America, and large portions of India and Asia. There will also be worldwide tropics.

The last time that the world was tropical from equator to near the poles was the Eocene of 55 million years ago. The spread of the tropics and new epeiric seas will radically change the distribution of tropical diseases. Both microbes and organismal vectors such as *Anopheles* mosquitoes will spread poleward. Malaria and dengue fever will be the major beneficiaries of these new ranges, but even rare diseases such as ebola will have far higher distribution. Because of the epeiric seas, rainfall on Earth will increase in previously dry areas, and thus we can expect mosquitoes, human parasites (such as those that cause elephantiasis and Africa sleeping sickness [*Trypanosoma*]), and round worms to vastly increase.

Finally, some millennia in the future, the Earth will again undergo the ultimate effect of global warming: the slowing and then cessation of the thermohaline distribution system. Because warm seawater holds less carbon dioxide and methane, there is a strong possibility that there will be major releases of these gases out of the oceans. If such release were to happen, we could expect very rapid warming, even faster than now, and this may trigger not only a mass extinction, but also a Permian-scale mass extinction.

All of the above is predicated on humans not stopping CO_2 rise. Getting around this will require some concerted global action. My colleague David Battisti believes that it will take a mass mortality of humans before our species gets its act together to do anything. The engineering required involves cutting carbon emissions, especially in all sectors of transportation, and bringing low carbon-producing energy facilities (which will inevitably be nuclear) online. Even so, these measures may not be enough.

A long-term solution, which has been posed by various scientific think tanks, is orbiting Sun shields, large sunshades that would reduce sunlight onto the planet. These could be positioned over the deep ocean so as to mitigate the problems of reduced plant growth under the shades.

HUMAN GREENHOUSE GAS PRODUCTION—MEDEAN EFFECT OR INCREASE IN GLOBAL BIOMASS?

It is valid to ask if a globally warmed world in the near future would in fact be a world with a greater biomass than now. If so, and with warmer temperatures and higher CO_2 levels, all models suggest that plants should grow faster and larger, and that even plankton should be at higher biomass levels. This, then, would be a Gaian effect— through the release of greenhouse gases, humans would, in fact, have made the Earth more habitable for more organisms than before. However, I do not think this is what will happen. The amount of new plant biomass would have to be balanced by the amount of biomass no longer produced because of global sea-level rise. Coastal forests and land plant regions normally contain much higher biomass than all but a few marine communities (coral reefs and eel grass flats, for example). The land areas flooded would be enormous. Second, because of the loss of land, large swaths of currently forested areas would have to be turned over to new cropland, and since any cropland lies fallow for one or more seasons each year, these new agricultural areas would have much lower biomass than the areas they replaced. Finally, the rise of sea level coming at a time when human population is cresting is bound to produce human conflict.

War reduces far more than human biomass: areas in combat zones are generally, sooner or later, "scorched earth" of one kind or another.

ENGINEERING SOLUTIONS

What are our choices? We simply cannot let the ice caps melt. To avoid that we need to reduce global temperatures, and we may have to do that with engineering if society does not have the will or ability to do it through conservation. Two solutions have been proposed: the first would be a series of large space mirrors, but there are no details on their construction or their cost. In 2005, however, Nobel Laureate Paul Crutzen proposed that injecting massive amounts of sulfur aerosols into the atmosphere, analogous to the effects of a large volcano but on a larger scale, would do the trick. The environmental side effects of this massive chemistry experiment, however, remain unknown, as Crutzen freely acknowledges.

Another solution might be covering large areas of the land, or sea, with reflective material. With a higher albedo, the temperature of the planet should drop.

In the long run, however, the engineering challenge will be getting carbon back into the atmosphere. Even with an enlarging Sun, the long-term drop in CO_2 as it is put into storage within continental rock, poses the most significant threat to planetary biomass. No plants means no oxygen, so we will require ever present efforts to move carbon from limestones and other continental rocks back into the atmosphere. This is relatively simple, as we know now—burn hydrocarbons. But as these will ultimately be used up, some kind of heating of limestones on a massive scale will do the trick.

[handwritten marginalia: in 1/2 billion yrs?]

EGGS IN THE BASKET

The old adage, do not keep all your eggs in one basket, is all too true. We humans should not be keeping our entire DNA "eggs" in basket Earth. But do we just salt away a lot of human eggs and sperm, or actually find another nest beyond the Earth?

Today there are no more uninhabited places on Earth that could sustain human life. That may change—the melting of the high lati-

[handwritten marginalia: Silly speculation nothing new]

tudes may see Antarctica and parts of Greenland and Siberia fit to hold more humans. But they are still on the Earth.

It is already thousands of years ago that we reached every corner of the planet, and some decades ago that we began planning the logical extension of our seemingly manifest destiny: the human migration off the Earth into space. With our first steps on the Moon, and now poised for a manned landing on Mars in this century, those genes pushing us ever onward and over the next hill remain dominant. Surely it is only a matter of time before we spread throughout our Milky Way Galaxy, and that message is such common knowledge in every corner of human society that it has become a cultural trope: spaceships, faster than light, that drive the ability to jump through space to far distant stars are so familiar that a large proportion of humanity believes that we either can already travel in this way or soon will. Ask any room filled with people if they believe in alien life, and more than half will answer yes. But ask the same room if they think that we will travel to the stars and the affirmative is virtually universal. Everyone has seen this interstellar travel countless times on television or on the silver screen, or read about it in magazines and books, some fiction and some not. Something so confidently and universally portrayed must be based on a coming reality. Or is it?

The obvious first questions are ones from technology and engineering: can a space vehicle be built that can take us—many of us at the same time—not only to Mars, but to the more distant stars? This is the point at which most discussions about space colonization both start and end. But we are dealing with something far more primal than the need to build warp drives. The prospects for colonization in space hinge on many biological and even sociological questions, as much as they rely on the purely technical, hardware sides of such a voyage. If engineering dictates the length of the journey, how will we know how many people to take, and what other organisms to take with us? How many worlds might we expect to be already suitable for human habitation without the need for extensive "terraforming" either in our solar system or among the hundred nearest star systems? But even more important, will the causes that have sent us on previous colonization voyages to the ends of the Earth now send us into space? Will some new imperative thrust us starward, or will our species stand at the edge of the void and turn back to spend human eternity on our home world? Even if we decide to go, is there any place that we can get to that is at all habitable for humanity—and can we get there with enough people for a colony to succeed?

When we humans began our conquest of planet Earth, we did so without worrying if the air was breathable over the next hill. The main obstacle to migration, at least where large lakes or seas blocked our way, was access to the right technology (in this case boats and ships), just as it remains a challenge to space migration. But at least there was no need for oxygen masks, or (at least for most regions on the Earth) special suits to help protect against the brutal cold of most planetary bodies beyond the Earth. Here we will look at the habitats in our own solar system, which could conceivably host humans. While there is the possibility of life on Mars, Europa, and Titan (the latter are moons of Jupiter and Saturn, respectively), it is really only Mars that might be habitable for humans over the long term. But how habitable? While many organizations, such as Robert Zubrin's Mars Society, maintain that Mars could be rendered habitable for humans (where one could breathe with only a minimum of technology for oxygen enrichment), the challenges are surely underestimated. While domed cities and even habitation of asteroids might be technologically feasible, the economics and reality are that it would be far easier to transport vast human populations to Antarctica than it would be to Mars, and probably just as useless. Mars has no plant life, and because it lacks plate tectonics, no mineral wealth. There would be very little that would drive the economy of a Martian colony.

There is another way to look at the Mars problem. It deals with financing the project. Consider the proposal for terraforming Mars by manufacturing halocarbon gases to cause a greenhouse warming of the planet, a plan floated by Zubrin and Wagner in 1996. Humans would make greenhouse gases on the surface of Mars, subsequent warming would cause the Martian soil to release its carbon dioxide, and genetically engineered plants would release oxygen from carbon

the great
unknown

dioxide. After nine hundred years of greenhouse warming, atmospheric pressure would increase to slightly less than the average atmospheric pressure in Denver or the normal cabin pressure in international air carriers. Humans who are acclimated to low atmospheric pressure might take up residence on Mars within seven hundred years. The price for this proposal is described as "several hundred billion dollars." But how would this money ever be recouped? Real estate sales would have to produce a staggering 1.36×10^{15} billion dollars to pay off the debt accumulated over seven hundred years. Thus an average square meter of Martian real estate would have to fetch 1,046 billion dollars to pay off the creditors. While we may hope for a vast, general increase in wealth over the next seven hundred years, this would still appear to make Martian real estate awfully pricey.

Mars may sustain small populations of scientists, but large human colonies on Mars may not be feasible. If that were the case, human colonization of the solar system would involve sealed cities in orbit.

What about the stars? Our Milky Way Galaxy is vast, composed of around 400 billion stars, a number that is seemingly inconceivable. It is a large "barred" spiral, and we have a pretty good idea about what our galaxy would look like if we could somehow view it from space. Our first impression would be of the sheer number of stars. But as large as this number is, the distances between these stars, in any common human measure of distance, is larger yet. Therefore, and assuming that some kind of spacecraft is developed that can make a journey over vast distances in space, what are our cosmic neighbors like? What kind of Milky Way neighborhood are we in? Is this a slum or the high-rent district, and, more important, are we in a high-density region where the stars are close together or are we far out in the country, separated from our nearest neighbors by vast distances?

How close is the nearest star? The nearest star to the Earth, apart from the Sun, is Proxima Centauri, which is 39.9 trillion kilometers or 4.2 light years. Thus light from Proxima Centauri takes 4.2 years to reach the Earth. If you took the French TGV, one of the fastest trains, using its highest recorded speed (515.3 kilometers per hour), on a trip to Proxima Centauri, it would take you about 8.86 million years. And from there they just keep getting farther away.

In discussions of interstellar travel, the focus is usually on the engineering challenges of building a vehicle that can travel between nearby stars. But the biology of humans and the animals, microbes, and plants taken by the would–be colonists might impose even greater challenges. Some sort of artificial gravity through ship spin might be necessary to allow successful human reproduction and development, and then there remain the challenges of human muscle and bone loss that come from long trips in space. Finally, what about some sort of deep sleep, suspended animation, or cryogenic freezing, after which, travelers would be revived upon arrival? Deep sleep actually involves significant short-term and long-term health risks, if it is possible at all. It looks like no one gets to doze off after take-off from the Earth and then wake up at Proxima, many years later, upon arrival at the Alpha Centauri star system.

SPACECRAFT

The long tradition of science fiction books, movies, and television shows has inculcated the belief that our species will be able to build a spaceship capable of rapidly (or even instantaneously) traveling between the stars. But the engineering challenges of building such a starship have been grossly underestimated.

The requirements of interstellar missions are beyond the performance of chemical propulsion, the type of rocket propulsion system used by all current space missions, even if accompanied by gravitational assist (where the spacecraft slings itself around a planet or the Sun to increase velocity). A good compromise between performance and technological availability can be found in solar sails, which would allow such missions to be performed at a limited cost and with limited technological studies. For a more distant future, such futuristic technology as "beamed energy sails"— where light energy from the Earth is focused on a retreating spacecraft's giant sail—and nuclear propulsion are both worth developing; the first would be appropriate for fast and smaller probes, while the second would be an enabling technology for a wide range of future space missions. But the problem with these systems is that they are very slow.

C'mon what is possible in a million or 10 or 100 million yrs!?

153

There is no doubt that the technology exists, and has long existed, for sending a human spacecraft to other star systems. Four spacecraft (the *Voyager* and *Pioneer* probes) are now traveling into interstellar space at speeds between 190 and 300 million miles per year (10.5 and 16.6 km/s). This performance was made possible by use of gravity assist: no missions beyond Mars orbit have been performed without it, and the lack of availability of a powerful enough rocket compelled us to exploit the gravity assist of Venus (twice) and of the Earth even for the Jupiter mission *Galileo*. Yet while such speed sounds really fast, in reality, these probes are traveling toward the stars at a veritable snail's pace.

So how long would the trip to a nearby star take with today's technology? Currently, the fastest spacecraft built can achieve a velocity of about 30 km per second (relative to the Earth). At that rate, the journey to Proxima Centauri would take about *40,000 years*! Additionally, at our current stage of space technology, the longest space missions that have been initiated are expected to have an operational lifetime of about forty years before failure of key components is likely to happen. Significant engineering advances such as automated self-repair may be required to ensure survival of any interstellar mission. In short, current spacecraft propulsion technology cannot send objects fast enough to reach the stars in a reasonable time.

Can any craft be built that will deliver humans within the maximum voyage durations listed in the preceding chapter? Chemical propulsion, characterized by low specific impulse but enabling engines with very large thrusts, falls short for deep-space and interstellar missions. Although the near interstellar space can be reached using chemical propulsion, aided by gravitational assist, no mission in interstellar space can be performed in a reasonable time without improvements in propulsion.

The two propulsion technologies that currently have the greatest potential must also overcome the greatest obstacles to be realized. Fusion engines, in which light elements combine to form heavier elements and energy, would release the fusion products as plasma from a magnetic nozzle. This would produce thrust with efficiency potentially as high as 250 times that of chemical rockets. However, controlled fusion has yet to live up to its potential. Similarly, antimatter engines hold the promise of amazing energy efficiency but are a long way off. Antimatter offers unrivaled energy density, as matter-antimatter annihilation releases the most energy per unit mass of any known physical reaction. The specific impulse from an antimatter engine could reach two hundred to two thousand times that of hydrogen/oxygen rockets, making antimatter the "hottest" potential propellant. However, the main obstacle at the moment is simply producing enough material; currently, antimatter costs $62.5 trillion per gram and can realistically be produced only in nanogram quantities. And it tends to blow up if it even touches normal matter. And we still have a problem with the amount of fuel needed.

So is the NEED!

The lightest mass U.S. manned spacecraft was the Mercury capsule—the *Liberty Bell*. It weighed only 2,836 pounds and launched on July 21, 1961. It would still take over *50 million kg* of antimatter fuel to get this tin can to the nearest star and back.

Surely breakthroughs will occur that somehow might radically increase speed or efficiency of a starship, or so goes the majority view when discussing this issue with the public. But the laws of physics are dispassionate. While it is true that upcoming breakthroughs are often unforeseen right up until their development (witness the rapid evolution of airplanes and cars after the first few years of their existence), there are virtually intractable problems facing engineers—such as trying to decelerate a starship that has achieved a significant fraction of the speed of light during an interstellar voyage. And there is no indication that any sort of propulsion system that even approaches the speed of light is even feasible.

and unstable!

of course

A LAST WORD

From the above, I have concluded that we are pretty much stuck on Earth. Of course breakthroughs may occur—compare technology of 1800 to the present day, for instance, or from A.D. 0 to the present day. Yet physics is physics. Building a flying machine in a rich and thick atmosphere was a lot easier than building a starship will be.

We are confined to the Earth, and, in the short term, our species will perhaps number in excess of 10 billion people. (For instance, on

For now—ok

physics of AD0 ≠ now; why would physics of 4000 be like today?

March 1, 2007, it was reported that the mean number of children for each female Rwandan of reproductive age is between five and six children). At the same time atmospheric carbon dioxide is near 400 ppm and steadily climbing. I estimate that 1,000 ppm will put us in lethal territory, for that figure will ensure the melting of all ice on land on our planet, which will bring on a slowing of ocean currents, followed by a greenhouse mass extinction.

Only engineering will save us now, for "nature" is simply the facts on the page, staring us in the face. Time to roll up the sleeves, take out the slide rules, encourage the boffins, and get to work. All of us.

REFERENCES

Introduction

Caldeira, K., and J. Kasting. 1992a. The life span of the biosphere revisited. *Nature* 360: 721–23.

———. 1992b. Susceptibility of the early Earth to irreversible glaciation caused by carbon ice clouds. *Nature* 359: 226–28.

Dole, S. 1964. *Habitable Planets for Man.* New York: Blaisdell.

Gott, J. 1993. Implications of the Copernican Principle for our future prospects. *Nature* 363: 315–19.

Gould, S. 1994. The evolution of life on Earth. *Scientific American* 271: 85–91.

Hart, M. 1979. Habitable zones around main sequence stars. *Icarus* 33: 23–39.

Kasting, J. 1996. Habitable zones around stars: An update. In *Circumstellar Habitable Zones*, ed. L. Doyle, 17–28. Menlo Park, CA: Travis House Publications.

Kasting, J., D. Whitmire, and R. Reynolds. 1993. Habitable zones around main sequence stars. *Icarus* 101: 108–28.

Laskar, J., F. Joutel, and P. Robutel. 1993. Stabilization of the Earth's obliquity by the Moon. *Nature* 361: 615–17.

Laskar, J., and P. Robutel. 1993. The chaotic obliquity of planets. *Nature* 361: 608–14.

McKay, C. 1996. Time for intelligence on other planets. In *Circumstellar Habitable Zones*, ed. L. Doyle, 405–19. Menlo Park, CA: Travis House Publications.

Schwartzman, D., and S. Shore. 1996. Biotically mediated surface cooling and habitability for complex life. In *Circumstellar Habit-*

able Zones, ed. L. Doyle, 421–43. Menlo Park, CA: Travis House Publications.

Volk, T. 1998. *Gaia's Body: Toward a Physiology of Earth*. New York: Copernicus.

Chapter 1

Bain, J. D., A. R. Chamberlin, C. Y. Switzer, and S. A. Benner. 1992. Ribosome-mediated incorporation of non-standard amino acids into a peptide through expansion of the genetic code. *Nature* 356: 537–39.

Bain, J. D., E. S. Diala, C. G. Glabe, T. A. Dix, and A. R. Chamberlin. 1989. Biosynthetic site-specific incorporation of a non-natural amino acid into a polypeptide. *J. Am. Chem. Soc.* 111: 8013–14.

Bains, W. 2001 The parts list of life. *Nat. Biotechnol.* 19: 401–2.

———. 2004. Many chemistries could be used to build living systems. *Astrobiology* 4: 137–67.

Baross, J. A., and J. W. Deming. 1995. Growth at high temperatures: Isolation and taxonomy, physiology, and ecology. In *The Microbiology of Deep-Sea Hydrothermal Vents*, ed. D. M. Karl, 169–217. Boca Raton, FL: CRC Press.

Benner, S. A. 1999. How small can a microorganism be? In *Size Limits of Very Small Microorganisms: Proceedings of a Workshop, Steering Group on Astrobiology of the Space Studies Board*, 126–35. Washington, DC: National Research Council.

———. 2004. Understanding nucleic acids using synthetic chemistry. *Accounts Chem. Res.* 37: 784–97.

Benner, S. A., and D. Hutter. 2002. Phosphates, DNA, and the search for nonTerran life. A second generation model for genetic molecules. *Bioorg. Chem.* 30: 62–80.

Benner, S. A., Ricardo, A., Carrigan, M. A. (2004) Is there a common chemical model for life in the universe? *Curr. Opinion Chem. Biol.* 8: 672–89.

Benner, S. A., and A. M. Sismour. 2005. Synthetic biology. *Nature Rev. Genetics* 6: 533–43.

Davies, Paul. 1998. *The Fifth Miracle*. Alan Lane/Penguin Press.

Dyson, F. 1999. *Origins of Life.* 2nd ed. Cambridge: Cambridge University Press.

Flechsig, E., and C. Weissmann. 2004. The role of PrP in health and disease. *Curr Mol Med.* 4: 337–53.

Haldane, J. 1947. *What Is Life?* New York: Boni and Gaer.

Jackson, G. S., and J. Collinge. 2001. The molecular pathology of CJD: Old and new variants. *J Clin Pathol: Mol Pathol.* 54: 393–99.

Luisi, P. L. 1998. About various definitions of life. *Origins Life Evol. Biosphere* 28: 613–22.

Olomucki, M. 1993. *The Chemistry of Life.* New York: McGraw-Hill.

Orgel, L. 1973. *The Origins of Life: Molecules and Natural Selection.* New York: Wiley.

Schrodinger, E. 1944. *What Is Life?* Cambridge: Cambridge University Press.

Serio, T. R., et al. 2001. Self-perpetuating changes in Sup35 protein conformation as a mechanism of heredity in yeast. *Biochem Soc Symp.* 68: 35–43.

Stetter, K. O. 1999. Extremophiles and their adaptation to hot environments. *FEBS Lett.* 452: 22–25.

———. 2002. Hyperthermophilic microorganisms. In *Astrobiology: The Quest for the Conditions of Life*, ed. G. Horneck and C. Baumstark-Khan, 169–84. Berlin: Springer.

Ward. P. 2005. *Life as We Do Not Know It.* New York: Viking Penguin.

West, R. 2002. Multiple bonds to silicon: 20 years later. *Polyhedron* 21: 467–72.

Zhang, L., A. Peritz, and E. Meggers. 2005. A simple glycol nucleic acid. *J. Am. Chem. Soc.* 127: 4174–75.

Chapters 2–4

Charlson, R. J., J. E. Lovelock, M. O. Andreae, and S. G. Warren. 1987. Oceanic phytoplankton, atmospheric sulphur, cloud albedo and climate. *Nature* 326: 655–61.

Cox, P. M., R. A. Betts, C. D. Jones, S. A. Spall, and I. J. Totterdell. 2000. Acceleration of global warming due to carbon-cycle feedbacks in a coupled climate model. *Nature* 408: 184–87.

Hamilton, W. D. 1995. Ecology in the large: Gaia and Ghengis Khan. *J. Appl. Ecol.* 32: 451–53.

Hardin, G. 1968. The tragedy of the commons. *Science* 162: 1243–48.

Holland, H. D. 1984. *The Chemical Evolution of the Atmosphere and Oceans.* Princeton, NJ: Princeton University Press.

Hutchinson, G. E. 1954. The biogeochemistry of the terrestrial atmosphere. In *The Earth as a Planet*, ed. G. P. Kuiper, 371–433. Chicago: University of Chicago Press.

Jahren, A. H., N. C. Arens, G. Sarmiento, J. Guerrero, and R. Amundson. 2001. Terrestrial record of methane hydrate dissociation in the early Cretaceous. *Geology* 29: 159–62.

Kamen, M. D. 1946. Survey of existing knowledge of biogeochemistry. 1. Isotopic phenomena in biogeochemistry. *Bull. Amer. Museum Nat. Hist.* 87: 110–38, 223–35.

Kirchner, J. W. 1990. Gaia metaphor unfalsifiable. *Nature* 345: 470.

———. 1991. The Gaia hypotheses: Are they testable? Are they useful? In *Scientists on Gaia*, ed. S. H. Schneider and P. J. Boston, 38–46. Cambridge: MIT Press.

———. 2002. The Gaia hypothesis: Fact, theory, and wishful thinking. *Clim. Change* 52: 391–408.

———. 2003. The Gaia Hypothesis: conjectures and refutations. *Climatic Change*, 58: 21–45.

Kirchner, J. W., and B. A. Roy. 1999. The evolutionary advantages of dying young: Epidemiological implications of longevity in metapopulations. *Amer. Nat.* 154: 140–59.

Kleidon, A. 2002. Testing the effect of life on Earth's functioning: How Gaian is the Earth System? *Clim. Change* 52: 383–89.

Lashof, D. A. 1989. The dynamic greenhouse: Feedback processes that may influence future concentrations of atmospheric trace gases in climatic change. *Clim. Change* 14: 213–42.

Lashof, D. A., B. J. DeAngelo, S. R. Saleska, and J. Harte. 1997. Terrestrial ecosystem feedbacks to global climate change. *Ann. Rev. Energy Environ.* 22: 75–118.

Legrand, M. R., R. J. Delmas, and R. J. Charlson. 1988. Climate forcing implications from Vostok ice-core sulphate data. *Nature* 334: 418–20.

Legrand, M., C. Feniet-Saigne, E. S. Saltzman, C. Germain, N. I. Barkov, and V. N. Petrov. 1991. Ice-core record of oceanic emissions of dimethylsulphide during the last climate cycle. *Nature* 350: 144–46.

Lenton, T. M. 1998. Gaia and natural selection. *Nature* 394: 439–47.

———. 2001. The role of land plants, phosphorus weathering and fire in the rise and regulation of atmospheric oxygen. *Global Change Biol.* 7: 613–29.

———. 2002. Testing Gaia: The effect of life on Earth's habitability and regulation. *Clim. Change* 52: 409–22.

Lenton, T. M., and W. von Bloh. 2001. Biotic feedback extends the life span of the biosphere. *Geophys. Res. Lett.* 28: 1715–18.

Lenton, T. M., and A. J. Watson. 2000. Redfield revisited 1. Regulation of nitrate, phosphate, and oxygen in the ocean. *Global Biogeochem. Cycles* 14: 225–48.

Lenton, T. M., and D. M. Wilkinson. 2003. Developing the Gaia theory: A response to the criticisms of Kirchner and Volk. *Clim. Change* 58.

Lovelock, J. E. 1983. Daisy world: A cybernetic proof of the Gaia hypothesis. *Coevolution Quarterly* 38: 66–72.

———. 1986. Geophysiology: A new look at earth science. In *The Geophysiology of Amazonia: Vegetation and Climate Interactions*, ed. R. E. Dickinson, 11–23. New York: Wiley.

———. 1988. *The Ages of Gaia.* New York: W. W. Norton.

———. 1990. Hands up for the Gaia hypothesis. *Nature* 344: 100–102.

———. 1991. *Gaia—The Practical Science of Planetary Medicine.* London: Gaia Books.

———. 1995. *The Ages of Gaia.* New York: W. W. Norton.

———. 2000. *Homage to Gaia: The Life of an Independent Scientist.* Oxford: Oxford University Press.

Lovelock, J. E., and L. R. Kump. 1994. Failure of climate regulation in a geophysiological model. *Nature* 369: 732–34.

Lovelock, J. E., and L. Margulis. 1974a. Homeostatic Tendencies of the Earth's Atmosphere. *Origins Life* 5: 93–103.

Lovelock, J. E., and L. Margulis. 1974b. Atmospheric homeostasis by and for the niosphere: The Gaia hypothesis. *Tellus* 26: 2–9.

Lovelock, J. E,. and A. J. Watson. 1982. The regulation of carbon dioxide and climate: Gaia or geochemistry. *Planet. Space Sci.* 30: 795–802.

Malthus, T. R. 1798. *An Essay on the Principle of Population as It Affects the Future Improvement of Society with Remarks on the Speculations of Mr. Godwin, M. Condorcet and Other Writers.* London: J. Johnson.

Petit, J. R., J. Jouzel, D. Raynaud, N. I. Barkov, J.-M. Barnola, I. Basile, M. Bender, J. Chappellaz, M. G. Delaygue, M. Delmotte, V. M. Kotlyakov, M. Legrand, V. Y. Lipenkov, C. Lorius, L. Pepin, C. Ritz, E. Saltzmank, M. and Stievenard. 1999. Climate and atmospheric history of the past 420,000 years from the Vostok ice core, Antarctica. *Nature* 399: 429–36.

Popper, K. R. 1963. *Conjectures and Refutations; The Growth of Scientific Knowledge.* London: Routledge and Kegan Paul.

Retallack, G. J. 2002. Carbon dioxide and climate over the past 300 myr. *Phil. Trans. Roy. Soc. London Series A* 360: 659–73.

Schimel, D. S., J. I. House, K. A. Hibbard, P. Bousquet, P. Ciais, P. Peylin, B. H. Braswell, M. J. Apps, D. Baker, A. Bondeau, J. Canadell, G. Churkina, W. Cramer, A. S. Denning, C. B. Field, P. Friedlingstein, C. Goodale, M. Heimann, R. A. Houghton, J. M. Melillo, B. Moore, D. Murdiyarso, I. Noble, S. W. Pacala, I. C. Prentice, M. R. Raupach, P. J. Rayner, R. J. Scholes, W. L. Steffen, and C. Wirth. 2001. Recent patterns and mechanisms of carbon exchange.

Schlesinger, W. H. 1997. *Biogeochemistry: An Analysis of Global Change.* San Diego: Academic Press.

Schneider, S. H. 2001. A goddess of Earth or the imagination of a man? *Science* 291: 1906–07.

Schwartzman, D., and C. H. Lineweaver. 2005, Temperature, Biogenesis, and Biospheric Self-Organization. In A. Kleidon, and R. Lorenz (eds.), *Non-equilibrium Thermodynamics and the Production of Entropy: Life, Earth and Beyond.* New York: Springer, 207–17.

Sober, E., and D. S. Wilson. 1998. *Unto Others: The Evolution and Psychology of Unselfish Behavior.* Cambridge: Harvard University Press.

Volk, T. 1998, *Gaia's Body: Toward a Physiology of Earth.* New York: Copernicus.

———. 2002. Toward a future for Gaia theory. *Clim. Change* 52: 423–30.

Watson, A. J., and P. S. Liss. 1998. Marine biological controls on climate via the carbon and sulphur geochemical cycles. *Phil. Trans. Roy. Soc. London, Series B* 353: 41–51.

Watson, A. J., and J. E. Lovelock. 1983. Biological homeostasis of the global environment: The parable of Daisyworld. *Tellus, Series B: Chem. Phys. Meterol.* 35: 284–89.

Woodwell, G. M., F. T. Mackenzie, R. A. Houghton, M. Apps, E. Gorham, and E. Davidson. 1998. Biotic feedbacks in the warming of the Earth. *Clim. Change* 40: 495–518.

Zachos, J., M. Pagani, L. Sloan, E. Thomas, and K. Billups. 2001. Trends, rhythms, and aberrations in global climate 65 ma to present. *Science* 292: 686–93.

Chapter 5

Berner, R. A. 1994. Geocarb II: A revised model of atmospheric CO_2 over Phanerozoic time. *Am. J. S.* 294: 56–91.

Kirchner, J. W. 1989. The Gaia hypothesis: Can it be tested? *Revue of Geophysics* 27: 223–35.

———. 1991. The Gaia hypotheses: Are they testable? Are they useful? In *Gaia: A New Look at Life on Earth*, ed. J. E. Lovelock, 38–46. Oxford: Oxford University Press., Oxford.

Lovelock, J. 1988. *The Ages of Gaia. A Biography of Our Living Earth.* New York: W. W. Norton.

———. 1992. A numerical model for biodiversity. *Phil. Trans. R. Soc. London B* 338: 383–91.

Lovelock, J. E., and S. R. Epton. 1975. The quest for Gaia. *New Scientist* 6.

Lovelock, J. E., and L. R. Kump. 1994. Failure of climate regulation in a geophysiological model. *Nature* 369: 732–34.

Lovelock, J. E., and L. Margulis. 1974a. Atmospheric homeostasis by and for the biophere: The Gaia hypothesis. *Tellus* 26: 1–10.

———. 1974b. Biological modulation of the Earth's atmosphere. *Icarus* 21: 471.

———. 1974c. Homeostatic tendencies of the Earth's atmosphere. *Origin of Life* 1: 12–22.

Lovelock, J. E., and A. J. Watson. 1982. The regulation of carbon dioxide and climate: Gaia or geochemistry. *Planet Space Science* 30: 795–802.

Sellers, A., and A. J. Meadows. 1975. Long-term variations in the albedo and surface temperature of the Earth. *Nature* 254: 44.

Schneider, S. H., and P. J. Boston, eds. 1991. *Scientists on Gaia.* Cambridge: MIT Press.

Vernadsky, V. 1945. The biosphere and the noosphere. *Amer. Sci.* 33: 1–12.

Watson, A. J., and J. E. Lovelock. 1983. Biological homeostasis of the global environment: The parable of Daisyworld. *Tellus* 35B: 284–89.

Chapters 6–7

Abe, Y., and T. Matsui. 1988. Evolution of an impact-generated H_2O-CO_2 atmosphere and formation of a hot proto-ocean on Earth. *J. Atmos. Sci.* 45: 3081–3101.

Armstrong, R. L. 1991. The persistent myth of crustal growth. *Aust. J. Earth Sci.* 38: 613–30.

Berner, R. A. 1991. A model for atmospheric CO_2 over Phanerozoic time. *Am. J. Sci.* 291: 339–76.

———. 1992. Weathering, plants, and long-term carbon cycle. *Geochim. Cosmochim. Acta* 56: 3225–31.

———. 1993. Paleozoic atmospheric CO_2: Importance of solar radiation and plant evolution. *Science* 261: 68–70.

———. 1994. Geocarb-II—A revised model of atmospheric CO_2 over Phanerozoic time. *Am. J. Sci.* 294: 56–91.

———. 1997. The rise of plants and their effect on weathering and atmospheric CO_2. *Science* 276: 544–46.

Berner, R.A., and D. M. Rye. 1992. Calculation of the Phanerozoic strontium isotope record of the ocean from a carbon cycle model. *Am. J. Sci.* 292: 136–48.

Berner, R.A., A. C. Lasaga, and R. M. Garrels. 1983. The carbonate-silicate geochemical cycle and its effect on atmospheric carbon dioxide over the past 100 million years. *Am. J. Sci.* 283: 641–83.

Bormann, B. T., D. Wang, F. H. Bormann, G. Benoit, R. April, and M. C. Snyder. 1998. Rapid, plant-induced weathering in an aggrading experimental ecosystem. *Biogeochemistry* 43: 129–55.

Caldeira, K., and J. F. Kasting. 1992. The life span of the biosphere revisited. *Nature* 360: 721–23.

Christensen, U. R. 1985. Thermal evolution models for the Earth. *J. Geophys. Res.* 90: 2995–3007.

Cochran, M. F., and R. A. Berner. 1992. The quantitative role of plants in weathering. In *Water-Rock Interaction*, ed. Y. K. Kharaka and A. S. Maest, 473–76.

Cohen, J. E. 1995. *How Many People Can the Earth Support?* New York: Norton.

Franck, S. 1992. Olivine flotation and crystallization of a global magma ocean. *Phys. Earth Planet. Inter.* 74: 23–28.

Franck, S., A. Block, W. von Bloh, C. Bounama, H. J. Schellnhuber, and Y. Svirezhev. 2000. Reduction of biosphere life span as a consequence of geodynamics. *Tellus* 52B: 94–107.

Franck, S., and Ch. Bounama 1995a. Effects of water-dependent creep rate on the volatile exchange between mantle and surface reservoirs. *Phys. Earth Planet. Inter.* 92: 57–65.

———. 1995b. Rheology and volatile exchange in the framework of planetary evolution. *Adv. Space. Res.* 15:79–86.

———. 1997. Continental growth and volatile exchange during Earth's evolution. *Phys. Earth Planet. Inter.* 100: 189–96.

Franck, S., and I. Orgzall. 1988. High-pressure melting of silicates and planetary evolution of Earth and Mars. *Gerlands Beitr. Geophys.* 97: 119–133.

François, L. M., and J.C.G. Walker. 1992. Modelling the phanerozoic carbon cycle and climate: constraints from the 87Sr/86Sr isotopic ratio of seawater. *Am. J. Sci.* 292: 81–135.

Goddéris, Y., and L. M. François. The cenozoic evolution of the strontium and carbon cycle: Relative importance of continental erosion and mantle exchanges. *Chem. Geology* 126: 169–90.

Gough, D. O. 1981. Solar interior structure and luminosity variations. *Sol. Phys.* 74: 21–34.

Henderson-Sellers, A., and B. Henderson-Sellers. 1988. Equable climate in the early Archaean. *Nature* 336: 117–18.

Kasting, J. F. 1982. Stability of ammonia in the primitive terrestrial atmosphere. *J. Geophys. Res.* 87: 3091–98.

———. 1984. Comments on the BLAG model: The carbonate-silicate geochemical cycle and its effect on atmospheric carbon dioxide over the past 100 million years. *Am. J. Sci.* 284: 1175–82.

———. 1987. Theoretical constraints on oxygen and carbon dioxide concentrations in the Precambrian atmosphere. *Precambrian Res.* 34: 205–29.

———. 1997. Warming early Earth and Mars. *Science* 276: 1213–15.

Kasting, J. F., S. M. Richardson, J. B. Pollack, and O. B. Toon. 1986. A hybrid model of the CO_2 geochemical cycle and its application to large impact events. *Am. J. S.* 286: 361–389.

Kuhn, W. R., J.C.G. Walker, and H. G. Marshall. 1989. The effect on Earth's surface temperature from variations in rotation rate, continent formation, solar luminosity, and carbon dioxide. *J. Geophys. Res.* 94: 11129–36.

Lasaga, A. C., R. A. Berner, R. M. Garrels. 1985. An improved geochemical model of atmospheric CO_2 fluctuations over past 100 million years. In *The Carbon Cycle and Atmospheric CO_2: Natural Variations Archaean to Present,* ed. E. T. Sundquist and W. S. Broecker, 397–411. Washington, DC: American Geophysical Union.

Lenton, T. M. 1998. Gaia and natural selection. *Nature* 394: 439–47.

Lenton, T. M., and W. von Bloh. 2001. Biotic feedback extends the life span of the biosphere. *Geophys. Res. Letters* 28: 1715–18.

Lovelock, J. E. 1995. *The Ages of Gaia—A Biography of Our Living Earth.* 2nd ed. Oxford: Oxford University Press.

Lovelock, J. E., and M. Whitfield. 1982. Life span of the biosphere. *Nature* 296: 561–63.

Marshall, H. G., J.C.G. Walker, and W. R. Kuhn. 1988. Long-term climate change and the geochemical cycle of carbon. *J. Geophys. Res.* 93: 781–801.

Matsui, T., and Y. Abe. 1986. Evolution of an impact-induced atmosphere and magma ocean on the accreting Earth. *Nature* 319: 303–5.

McGovern, P. J., and G. Schubert. 1989. Thermal evolution of the Earth: Effects of volatile exchange between atmosphere and interior. *Earth Planet Sci. Lett.* 96: 27–37.

Meissner, R. 1986. *The Continental Crust.* Orlando: Academic Press.

Moulton, K., and R. A. Berner. 1998. Quantification of the effect of plants on weathering: Studies in Iceland. *Geology* 26: 895–98.

Newson, H. E., and J. H. Jones, eds. 1990. *Origin of the Earth.* New York: Oxford University Press, and Houston: Lunar and Planetary Institute.

Owen, T., R. D. Cess, and V. Ramanathan. 1979. Enhanced CO_2 greenhouse to compensate for reduced solar luminosity on early Earth. *Nature* 277: 640–42.

Raup, D. M., and J. J. Sepkoski. 1982. Mass extinctions in the marine fossil record. *Science* 215: 1501–03.

Reymer, A., and G. Schubert. 1984. Phanerozoic addition rates of the continental crust and crustal growth. *Tectonics* 3: 63–67.

Rye, R., P. H. Kuo, and H. D. Holl. 1997. Atmospheric carbon dioxide concentrations before 2.2 billion years ago. *Nature* 378: 603–5.

Sagan, C. 1977. Reduced greenhouse and the temperature history of the Earth and Mars. *Nature* 269: 224–26.

Sagan, C. and Ch. Chyba. 1997. The early faint young Sun paradox: Organic shielding of ultraviolet-labile greenhouse gases. *Science* 276: 1217–21.

Sagan, C. and G. Mullen. 1972. Earth and Mars: Evolution of atmospheres and surface temperatures. *Science* 177: 52–56.

Schneider, S. H., and P. J. Boston, eds. 1993. *Scientists on Gaia.* Cambridge: MIT Press.

Schwartzman, D. W. 1999. *Life, Temperature and the Earth: The Self-organizing Biosphere.* New York: Columbia University Press.

Schwartzman, D. W., and T. Volk. 1989. Biotic enhancement of weathering and the habitability of Earth. *Nature* 340: 457–60.

Stevenson, D. J., T. Spohn, and G. Schubert. 1983. Magnetism and the thermal evolution of the terrestrial planets. *Icarus* 54: 466–89.

Stumm, W., and J. J. Morgan, eds. 1981. *Aquatic Chemistry.* New York: Wiley.

Tajika, E., and T. Matsui. 1992. Evolution of terrestrial proto-CO_2 atmosphere coupled with thermal history of the Earth. *Earth Planet. Sci. Lett.* 113: 251–66.

Taylor, S. R., and S. M. McLennan. 1995. The geological evolution of the continental crust. *Rev. Geophysics* 33: 241–65.

Turcotte, D. L., and G. Schubert. 1982. *Geodynamics.* New York: Wiley.

Volk, T. 1987. Feedbacks between weathering and atmospheric CO_2 over the last 100 million years. *Am. J. Sci.* 287: 763–79.

Walker, J.C.G. 1982. Climatic factors on the Archean Earth. *Palaeogeogr. Palaeoclimat. Palaeoecol.* 40: 1–11.

Walker, J.C.G., P. B. Hays, and J. F. Kasting. 1981. A negative feedback mechanism for the long-term stabilization of Earth's surface temperature. *J. Geophys. Res.* 86: 9776–82.

Watson, A. J., and J. E. Lovelock. 1983. Biological homeostasis of the global environment: The parable of Daisyworld. *Tellus* 35B: 284–89.

Williams, D. R., and V. Pan. 1992. Internally heated mantle convection and the thermal and degassing history of the Earth. *J. Geophys. Res.* 97 B6: 8937–50.

Zahnle, K. J., J. F. Kasting, and J. B. Pollack. 1988. Evolution of a steam atmosphere during Earth's formation. *Icarus* 74: 62–97.

Chapter 8

Brown, L, C. Flavin, and H. French. 1999. *State of the World, 1999.* New York: Norton/Worldwatch Books.

Cohen, J. 1995. *How Many People Can the Earth Support?* New York: W. W. Norton.

Fuller, E. 1987. *Extinct Birds.* New York: Facts on File.

Goudie, A., and H. Viles. 1997. *The Earth Transformed.* New York: Blackwell.

Hallam, A., and P. Wignall. 1997. *Mass Extinctions and Their Aftermath.* Oxford: Oxford University Press.

McKinney, M., ed. 1998. *Diversity Dynamics.* New York: Columbia University Press.

Salvadori, F. 1990. *Rare Animals of the World.* New York: Mallard Press.

Stanley, S. 1987. *Extinctions*. San Francisco: W. H. Freeman.

Walker, J.C.G. 1977. *Evolution of the Atmosphere*. London: Macmillan.

Ward, P. 1994. *The End of Evolution*. New York: Bantam Doubleday Dell.

Wilson, E. 1992. *The Diversity of Life*. Cambridge: Harvard University Press.

Chapters 10–11

Benarde, M. A. 1992. *Global Warning—Global Warming*. New York: Wiley.

Berner, R. A., and A. C. Lasaga. 1989. Modeling the geochemical carbon cycle. *Scientific American* 260: 74.

Bilger, B. 1992. *Global Warming*. New York: Chelsea House.

Bolin, B. 1986. *The Greenhouse Effect, Climatic Change, and Ecosystems*. New York: Wiley.

Cerling, T. E., J. R. Ehleringer, and J. M. Harris. 1998. Carbon dioxide starvation, the development of C_4 ecosystems, and mammalian evolution. *Phil. Trans. R. Soc. Lond. B* 353: 159–71.

Cerling, T. E., J. M. Harris, B. J. MacFadden, M. G. Leakey, J. Quade, V. Eisenmann, and J. R. Ehleringer. 1997. Global vegetation change through the Miocene-Pliocene boundary. *Nature* 389: 153–58.

Chapin, F. S. 1992. *Arctic Ecosystems in a Changing Climate: An Ecophysiological Perspective*. San Diego: Academic Press.

Clarkson, J., and J. Schmandt. 1992. *The Regions and Global Warming: Impacts and Response Strategies*. New York: Oxford University Press.

Condie, K. C. 1984. *Plate Tectonics and Crustal Evolution*. 2nd ed. Oxford: Pergamon Press.

Cotton, W. R., and R. A. Pielke. 1995. *Human Impacts on Weather and Climate*. New York: Cambridge University Press.

Coward, H. G., and T. Hurka. 1993. *The Greenhouse Effect: Ethics and Climate Change*. Waterloo: Wilfrid Laurier University Press.

Cox, A. 1973. *Plate Tectonics and Geomagnetic Reversals*. San Francisco: W. H. Freeman.

Crutzen, P. J., and T. E. Graedel. 1995. *Atmosphere, Climate, and Change*. New York: Scientific American Library.

Dalziel, I.W.D. 1992. On the organization of American plates in the Neoproterozoic and the breakout of Laurentia. *GSA Today* 2: 237.

DePaolo, D. J. 1984. The mean life of continents; estimates of continental recycling from Nd and Hf isotopic data and implications for mantle structure. *Geophys. Res. Lett.* 10: 705–8.

Dietz, R. S. 1961. Continent and ocean basin evolution by spreading of the sea floor. *Nature* 190: 854–57.

Dornbusch, R., and J. Poterba. 1991. *Global Warming*. Cambridge: MIT Press.

Ehleringer, J. R., and T. E. Cerling. 1995. Atmospheric CO_2 and the ratio of intercellular to ambient CO_2 levels in plants. *Tree Physiol.* 15: 105–11.

Ehleringer, J. R., T. E. Cerling, and B. R. Helliker. 1997. C_4 photosynthesis, atmospheric CO_2, and climate. *Oecologia* 112: 285–99.

Ephraums, J. J., J. T. Houghton, and G. J. Jenkins. 1990. *Climate Change: The IPCC Scientific Assessment*. New York: Cambridge University Press.

Firor, J. 1990. *The Changing Atmosphere: A Global Challenge*. New Haven: Yale University Press.

Fisher, D. E. 1990. *Fire and Ice: The Greenhouse Effect, Ozone Depletion, and Nuclear Winter*. New York: Harper and Row.

Flavin, C. 1989. *Slowing Global Warming: A Worldwide Strategy*. Washington, DC: Worldwatch Institute.

Fyfe, W. S. 1978. The evolution of the Earth's crust: Modern plate tectonic to ancient hotspot tectonic? *Chem. Geol.* 23: 89–114.

Gates, D. M. 1993. *Climate Change and Its Biological Consequences*. Sunderland, MA: Sinauer Associates.

Gay, K. 1986. *The Greenhouse Effect*. New York: F. Watts.

Glantz, M. H. 1991. The use of analogies in forecasting ecological and societal responses to global warming. *Environment* 33: 10.

Gribbin, J. R. 1982. *Future Weather and the Greenhouse Effect*. New York: Delacorte Press/Eleanor Friede.

———. 1990. *Hothouse Earth: The Greenhouse Effect and Gaia*. New York: Grove Weidenfeld.

Hess, H. H. 1962. History of ocean basins. In *Petrologic Studies—a Volume to Honor A. F. Buddington*, ed. A.E.J. Engel et al., 599–620. Boulder: Geological Society of America.

Hare, T., and A. Khan. 1990. *The Greenhouse Effect*. New York: Gloucester Press.

Hewitt, C. N., and W. T. Sturges. 1993. *Global Atmospheric Chemical Change*. New York: Elsevier Applied Science.

Jones, P. D., and T.M.L. Wigley. 1990. Global warming trends. *Scientific American* 263: 84.

Kellogg, W.W. and R. Schware. *Climate Change and Society: Consequences of Increasing Atmospheric Carbon Dioxide*. Boulder: Westview Press.

Krause, F. 1992. *Energy Policy in the Greenhouse*. New York: Wiley.

Lovejoy, T. E., and R. L. Peters. 1992. *Global Warming and Biological Diversity*. New Haven: Yale University Press.

McCuen, G. E. 1987. *Our Endangered Atmosphere: Global Warming and the Ozone Layer*. Hudson: Gary E. McCuen Publications.

Mesirow, L. E., and S. H. Schneider. 1976. *The Genesis Strategy: Climate and Global Survival*. New York: Plenum Press.

Mitchell, G. J. 1991. *World on Fire: Saving an Endangered Earth*. Toronto: Collier Macmillan Canada.

Nance, J. J. 1991. *What Goes Up: The Global Assault on Our Atmosphere*. New York: W. Morrow.

Oppenheimer, M. 1990. *Dead Heat: The Race against the Greenhouse Effect*. New York: Basic Books.

Reckman, A. 1991. *Global Warming*. New York: Gloucester Press.

Reukin, A. 1992. *Global Warming: Understanding the Forecast*. New York: Abbeville Press.

Schneider, S. H. 1989. *Global Warming: Are We Entering the Greenhouse Century?* San Francisco: Sierra Club Books.

———. 1990a. Debating GAIA. *Environment* 32: 4.

———. 1990b. Prudent planning for a warmer planet. *New Scientist* 128: 49.

Sedjo, R. A. 1989. Forests: A tool to moderate global warming. *Environment* 31: 14.

Simons, P. 1992. Why global warming could take Britain by storm. *New Scientist* 136: 35.

South, E. L. 1990. *The Changing Atmosphere: A Global Challenge*. New Haven: Yale University Press.

Uyeda, S. (1987) *The New View of the Earth*. San Francisco: W. H. Freeman.

Vine, F. J., and D. H. Mathews. 1963. Magnetic anomalies over oceanic ridges. *Nature* 199: 947–49.

Wegener, A. 1912. Die Entstehung der Kontinente. *Geol. Rundschau* 3:276–92.

———. 1924. *The Origin of Continents and Oceans*. London: Methuen.

Wilson, J. T. 1965. A new class of faults and their bearing on continental drift. *Nature* 207: 343–46.

INDEX